Open Problems in the Geometry and Analysis
of Banach Spaces

Antonio J. Guirao • Vicente Montesinos
Václav Zizler

Open Problems in the Geometry and Analysis of Banach Spaces

 Springer

Antonio J. Guirao
Departamento de Matemática Aplicada
Instituto de Matemática Pura y Aplicada
Universitat Politècnica de València
Valencia, Spain

Vicente Montesinos
Departamento de Matemática Aplicada
Instituto de Matemática Pura y Aplicada
Universitat Politècnica de València
Valencia, Spain

Václav Zizler
Department of Mathematical and Statistical
  Sciences
University of Alberta
Alberta, Canada

ISBN 978-3-319-81551-0   ISBN 978-3-319-33572-8   (eBook)
DOI 10.1007/978-3-319-33572-8

Printed on acid-free paper

This Springer imprint is published by Springer Nature
The registered company is Springer International Publishing AG Switzerland

Dedicated to the memory of
Joram Lindenstrauss,
Aleksander Pełczyński,
and Manuel Valdivia

Preface

This is a commented collection of some easily formulated open problems in the geometry and analysis on Banach spaces, focusing on basic linear geometry, convexity, approximation, optimization, differentiability, renormings, weak compact generating, Schauder bases and biorthogonal systems, fixed points, topology, and nonlinear geometry.

The collection consists of some commented questions that, to our best knowledge, are open. In some cases, we associate the problem with the person we first learned it from. We apologize if it may turn out that this person was not the original source. If we took the problem from a recent book, instead of referring to the author of the problem, we sometimes refer to that bibliographic source. We apologize that some problems might have already been solved. Some of the problems are long-standing open problems, some are recent, some are more important, and some are only "local" problems. Some would require new ideas, and some may go only with a subtle combination of known facts. All of them document the need for further research in this area. The list has of course been influenced by our limited knowledge of such a large field. The text bears no intentions to be systematic or exhaustive. In fact, big parts of important areas are missing: for example, many results in local theory of spaces (i.e., structures of finite-dimensional subspaces), more results in Haar measures and their relatives, etc. With each problem, we tried to provide some information where more on the particular problem can be found. We hope that the list may help in showing borders of the present knowledge in some parts of Banach space theory and thus be of some assistance in preparing MSc and PhD theses in this area. We are sure that the readers will have no difficulty to consider as well problems related to the ones presented here. We believe that this survey can especially help researchers that are outside the centers of Banach space theory. We have tried to choose such open problems that may attract readers' attention to areas surrounding them.

Summing up, the main purpose of this work is to help in convincing young researchers in functional analysis that the theory of Banach spaces is a fertile field of research, full of interesting open problems. Inside the Banach space area, the text should help a young researcher to choose his/her favorite part to work in. This way we hope that problems around the ones listed below may help in motivating research in these areas. For plenty of open problems, we refer also to, e.g., [AlKal06, BenLin00, BorVan10, CasGon97, DeGoZi93, Fa97, FHHMZ11, FMZ06, HaJo14, HMZ12, HMVZ08, HajZiz06, Kal08, LinTza77, MOTV09, Piet09, Woj91], and [Ziz03].

To assist the reader, we provided two indices and a comprehensive table referring to the listed problems by subject.

We follow the basic notation in the Banach space theory and assume that the reader is familiar with the very basic concepts and results in Banach spaces (see, e.g., [AlKal06, Di84, FHHMZ11, LinTza77, HMVZ08, Megg98]). For a very basic introduction to Banach space theory—"undergraduate style"—we refer to, e.g., [MZZ15, Chap. 11]. By a **Banach space** we usually mean an infinite-dimensional Banach space over the real field—otherwise we shall spell out that we deal with the finite-dimensional case. If no confusion may arise, the word **space** will refer to a Banach space. Unless stated otherwise, by a **subspace** we shall mean a closed subspace. The term **operator** refers to a bounded linear operator, and an operator with real values will be called a **functional**, understanding, except if it is explicitly mentioned, that it is continuous. A subspace Y of a Banach space X is said to be **complemented** if it is the range of a bounded linear projection on X. The **unit sphere** of the Banach space X, $\{x \in X : \|x\| = 1\}$, is denoted by S_X, and the **unit ball** $\{x \in X : \|x\| \le 1\}$ is denoted by B_X. The words "smooth" and "differentiable" have the same meaning here. Unless stated otherwise, they are meant in the Fréchet (i.e., total differential) sense. If they are meant in the Gâteaux (i.e., directional) sense, we clearly mention it (for those concepts, see their definitions in, e.g., [FHHMZ11, p. 331]). We say that a norm is **smooth** when it is smooth at all nonzero points. Sometimes, we say that "a Banach space X admits a norm $\| \cdot \|$," meaning that it admits an *equivalent* norm $\| \cdot \|$. By **ZFC** we mean, as usual, the Zermelo-Fraenkel-Choice standard axioms of set theory. Unless stated otherwise, we use this set of axioms. We say that some statement is **consistent** if its negation cannot be proved by the sole ZFC set of axioms. Cardinal numbers are usually denoted by $\aleph$, while ordinal numbers are denoted by α, β, etc. With the symbol $\aleph_0$ we denote the cardinal number of the set $\mathbb{N}$ of natural numbers, and $\aleph_1$ is the first uncountable cardinal. Similarly, ω_0 (sometimes denoted ω) is the ordinal number of the set $\mathbb{N}$ under its natural ordering, and ω_1 is the first uncountable ordinal. The continuum hypothesis then reads $2^{\aleph_0} = \aleph_1$. The cardinality $2^{\aleph_0}$ of the set of real numbers (the **continuum**) will be denoted by c. If no confusion may arise, we sometimes denote by ω_1 also its cardinal number $\aleph_1$.

We prepared this little book as a working companion for [FHHMZ11] and [HMVZ08]. We often use this book to upgrade and update information provided in these two references.

Overall, we would be glad if this text helped in providing a picture of the present state of the art in this part of Banach space theory. We hope that the text may serve also as a kind of reference book for this area of research.

Acknowledgments The third named author would like to express his gratitude to the late Joram Lindenstrauss and Olek Pełczyński for their lifelong moral support and encouragement to the Prague Banach space group, in particular in connection with the organization of Prague annual international winter schools for young researchers. In fact, for the young, starting Prague group, the moral support of the Israeli and Polish schools in the seventies and eighties of the last century was vital. The second named author is grateful for Olek's encouraging attitude regarding mathematics and his personal friendship and the first and second named authors to the Prague group for its continued support, encouragement, and friendship. All three authors want to dedicate this work also to the founder of the modern Spanish functional analysis school, the late Manuel Valdivia.

The authors thank the Springer team, especially Marc Strauss, for their interest in this text and Mr. Saswat, Mishra, for his professional editing of the manuscript. The authors thank their colleagues who helped by various means, advice or references, etc., to this text.

The third named author appreciates the electronic access to the library of the University of Alberta. The first two named authors want also to thank the Universitat Politècnica de València, its Instituto de Matemática Pura y Aplicada, and its Departamento de Matemática Aplicada, for their support and the working conditions provided. The authors were also supported in part by grants MTM2011-22417 and MICINN and FEDER (Project MTM2014-57838-C2-2-P) (Vicente Montesinos) and Fundación Séneca (Project 19368/PI/14), and MICINN and FEDER (Project MTM2014-57838-C2-1-P) (Antonio J. Guirao).

The material comes from the interaction with many colleagues in meetings, in work, and in private conversations and, as the reader may appreciate in the comments to the problems, from many printed sources—papers, books, reviews, and even beamers from presentations—and, last, from our own research work. It is clear then that it will be impossible to explicitly thank so many influences. The authors prefer to carry on their own shoulders the responsibility for the selection of problems, eventual inaccuracies, wrong attributions, or lack of information about solutions. The names of authors appearing in problems, in comments, and in the reference list correspond to the panoply of mathematicians to whom thanks and acknowledgment usually appear in the introduction to a book.

Above all, the authors are indebted to their families for their moral support and encouragement.

The authors wish the readers a pleasant time spent over this little book.

Valencia, Spain   A.J. Guirao
Valencia, Spain   V. Montesinos
Calgary, Canada   V. Zizler
2016

Contents

Chapter 1
Basic Linear Structure

1.1 Schauder Bases

A sequence $\{e_i\}_{i=1}^{\infty}$ in a Banach space X is called a **Schauder basis for** X if for each $x \in X$ there is a unique sequence of scalars $\{\alpha_i\}_{i=1}^{\infty}$ such that $x = \sum_{i=1}^{\infty} \alpha_i e_i$. If the convergence of this series is **unconditional** for all $x \in X$ (i.e., any rearrangement of it converges), we say that the Schauder basis is **unconditional**. This is equivalent to say that under any permutation $\pi : \mathbb{N} \to \mathbb{N}$, the sequence $\{e_{\pi(i)}\}_{i=1}^{\infty}$ is again a basis of X.

*Not every separable Banach space admits a Schauder basis,*as it was shown first by P. Enflo (see, e.g., [LinTza77, p. 29] or [FHHMZ11, p. 711]).

We refer to [DLAT10, Pe06] and [Cass01] for more on the questions formulated in Problem 1.

Problem 1. Let X be a separable infinite-dimensional Banach space that is not isomorphic to a Hilbert space.

  (i) (A. Pełczyński) Does there exist an infinite-dimensional subspace of X with Schauder basis that is not complemented in X?

  (ii) (A. Pełczyński) Do there exist two infinite-dimensional subspaces of X with Schauder basis that are not isomorphic?

  (iii) (A. Pełczyński) Does there exist a subspace of X with Schauder basis that is not isomorphic to ℓ_2?

(continued)

© Springer International Publishing Switzerland 2016
A.J. Guirao et al., *Open Problems in the Geometry and Analysis of Banach Spaces*,
DOI 10.1007/978-3-319-33572-8_1

Problem 1 (continued)

(iv) (G. Godefroy) Does there exist an infinite sequence $\{X_n\}_{n=1}^{\infty}$ of mutually nonisomorphic infinite-dimensional subspaces of X?

(v) Does there exist a subspace of X that has no unconditional Schauder basis?

Concerning Problem 1 we would like to point out:

1. J. Lindenstrauss and L. Tzafriri showed in [LinTza71] that *if all subspaces of a Banach space X are complemented in X, then X is isomorphic to a Hilbert space* (cf., e.g., [FHHMZ11, p. 309] or [AlKal06, p. 301]). On the other hand, W. B. Johnson and A. Szankowski in [JoSz14] showed that *there is a separable infinite-dimensional Banach space X that is not isomorphic to a Hilbert space and yet, every subspace of X is isomorphic to a complemented subspace of X.*

2. R. Komorowski and N. Tomczak-Jaegermann [KoTo95,98] and W. T. Gowers [Gow96] showed that *a separable Banach space X is isomorphic to a Hilbert space if all infinite-dimensional subspaces of X are isomorphic to X* (cf., e.g., [FHHMZ11, p. 267]).

3. W. B. Johnson showed that *there is a separable reflexive infinite-dimensional Banach space with unconditional Schauder basis* (the so-called 2-**convexified Tsirelson space** T^2) *that does not contain isomorphic copies of ℓ_p, $1 < p < \infty$, and such that all of its subspaces do have Schauder basis* (cf., e.g., [Cass01, p. 276]).

4. A. Pełczyński and I. Singer showed in [PeSi64] that *if an infinite-dimensional Banach space X has a Schauder basis, then there is a continuum of normalized mutually non-equivalent conditional Schauder bases in X.* A Schauder basis $\{e_i\}_{i=1}^{\infty}$ is **normalized** if $\|e_i\| = 1$ for all i and a basis $\{e_i\}_{i=1}^{\infty}$ is **equivalent to a basis** $\{f_i\}_{i=1}^{\infty}$ if for scalars $\{\lambda_i\}_{i=1}^{\infty}$, $\sum \lambda_i e_i$ converges if and only if $\sum \lambda_i f_i$ converges.

5. R. Anisca *verified in* [An10] *in the positive G. Godefroy conjecture in Problem 1 (iv) for the class of the so-called weak Hilbert spaces that are not isomorphic to Hilbert spaces.* A Banach space X is a **weak Hilbert space** if there are positive constants K and δ such that every n-dimensional subspace of X has a subspace of dimension at least δn that is K-linearly isomorphic to a Hilbert space and K-complemented in X. The spaces X and Y are K-**linearly isomorphic** if there is a linear isomorphism φ from X onto Y such that $\|\varphi\| \cdot \|\varphi^{-1}\| \leq K$. The space Y is K-**complemented in** X if there is a projection P from X onto Y such that $\|P\| \leq K$. All subspaces, quotients, and duals of weak Hilbert spaces are themselves weak Hilbert spaces, and W. B. Johnson showed that *all weak Hilbert spaces are superreflexive* (for references see, e.g., [Pis88]). **Superreflexive spaces** are spaces that admit an equivalent uniformly convex norm. The norm $\| \cdot \|$ of a

Banach space is said to be **uniformly convex** if the **modulus of convexity** $\delta(\varepsilon) := \inf\{1 - \|(x + y)/2\| : x, y \in B_X, \|x - y\| \geq \varepsilon\}$ is positive for every $\varepsilon \in (0, 2]$.

6. A sequence in a normed space is said to be **basic** if it is a Schauder basis of its closed linear span. *Every infinite-dimensional Banach space contains a basic sequence* (a classic result of S. Mazur). W. T. Gowers and B. Maurey found, in 1991, *a reflexive Banach space X that has no unconditional basic sequence* (i.e., *X has no infinite-dimensional subspace with unconditional basis*). The result appeared in [GowMau93], and was inspired, as B. Maurey indicates in [Mau03], by the famous **B. S. Tsirelson example** of *a reflexive Banach space that does not contain any ℓ_p for $1 < p < +\infty$* [Tsi74] and by its modification by Th. Schlumprecht [Schl91]. It was crucial that W. B. Johnson showed that *this space of W. T. Gowers and B. Maurey is actually* **hereditarily indecomposable** (**HI**, in short), i.e., a Banach space X such that no closed subspace Z of X can be written as a topological direct sum of two infinite-dimensional closed subspaces of Z. This means that for every pair of two closed infinite-dimensional subspaces Y and Z of such X, the distance of S_Y to S_Z is zero. This in turn means that in such X, there is no bounded projection P from a subspace Z into itself such that the range of P and the kernel of P were infinite-dimensional (see also Sect. 1.2 below). Note that *all HI spaces clearly have the property that they do not contain any unconditional basic sequence*—since the span of such a sequence would be clearly decomposable. It follows that *if X is a hereditarily indecomposable space, then X is not isomorphic to any proper subspace of X; in particular, it is not isomorphic to any of its hyperplanes.* This was an open problem from S. Banach himself (cf., e.g., [Mau03, p. 1265]). The first example of a Banach space not isomorphic to its hyperplanes was found by W.T. Gowers in [Gow94]. This space has an unconditional basis. See also Problem 2 and Remarks to it, as well as Sect. 1.2 below.

These results solved problems that have stayed open for about 70 years and created a true revolution in the recent development of Banach space theory. For this and more information we recommend to consult, e.g., [Mau03].

Let us note in passing that *there is a nonseparable C(K)-space such that every one-to-one operator from C(K) into itself is necessarily onto* [AviKo13].

W. T. Gowers proved in [Gow96] the following dichotomy: *Let X be an arbitrary infinite-dimensional Banach space. Either X contains an unconditional basic sequence or X contains a HI subspace.* He also produced *a Banach space Y not containing any reflexive infinite-dimensional subspace and containing no copies of c_0 or ℓ_1* [Gow94b]. By James' theorem (cf., e.g., [FHHMZ11, p. 204]), together with the dichotomy just mentioned, we get a HI subspace of Y that has no reflexive subspace.

In this direction see also [Fer97].

A **bump function** (or just a **bump**) on a Banach space is a real-valued function with bounded nonempty support. R. Deville showed that *if a Banach space X admits a C^∞-smooth bump function, then X contains a copy of c_0 or some ℓ_p, $p > 1$*, so it cannot be hereditarily indecomposable (see Remark 6 to Problem 1; see also [DeGoZi93, p. 209]).

We do not know the answer to the following problem:

Problem 2. Assume that X is a separable infinite-dimensional Banach space that admits a C^∞-smooth bump function. Is X necessarily isomorphic to its hyperplanes?

1. The very original space T of Tsirelson (see Remark 6 to Problem 1 above) was constructed as *a reflexive space with unconditional basis, no infinite-dimensional subspace of which admits a uniformly convex norm*. This short self-contained crystal-clear text has drastically influenced the whole Banach space theory. This was achieved by ensuring that *for every infinite-dimensional subspace E of T, c_0 is **crudely finitely representable in** E* (meaning that there is $K > 0$ such that every finite-dimensional subspace of c_0 is K-isomorphic to a subspace of E). So, c_0 is crudely finitely representable in each infinite-dimensional subspace of T and yet, c_0 is not isomorphic to any subspace of T (T is reflexive). Now, if an infinite-dimensional subspace E of T had an equivalent uniformly convex norm, by a simple limit technique explained, e.g., in [FHHMZ11, p. 435], this would give that c_0 admits an equivalent uniformly convex norm. Since c_0 does not admit any uniformly convex norm as it is not reflexive [FHHMZ11, p. 434], this all implies that no infinite-dimensional subspace of T can have an equivalent uniformly convex norm. Thus, in particular, no ℓ_p for $p > 1$ can be isomorphic to a subspace of T. As a reflexive space, T cannot contain an isomorphic copy of ℓ_1. Therefore T cannot contain a copy of any ℓ_p or c_0. Tsirelson's original, truly ingenious, short, direct geometric construction of the unit ball of T [Tsi74] is described, e.g., in [FHHMZ11, p. 459]. The key point is the construction of the unit ball of T as a weakly compact subset of c_0, in such a way that, by Pełczyński method, one can model finitely c_0 on the "tails" of sequences. This kind of modelling is the main novelty in Tsirelson construction. It is proved in [CJT84] that *T isomorphically embeds into each infinite-dimensional subspace of T*. An analytic approach to the Tsirelson (dual) space is explained in [LinTza77, p. 95]. This space thus solved the original Banach problem on containment of ℓ_p or c_0 in every Banach space which used to be a famous longstanding problem for about 40 years. Tsirelson's example had an enormous impact on the Banach space theory and has been substantially influencing its further development since the year 1974, when it appeared. The reader is encouraged to consult [CaSh89].

As a reflexive space, Tsirelson's space admits a Fréchet differentiable norm (cf., e.g., [FHHMZ11, p. 387]). However, the (continuous) differential of this norm cannot be locally uniformly continuous on (the sphere of) any infinite-dimensional subspace of the Tsirelson space (cf., e.g., [DeGoZi93, p. 203]).

A space introduced by T. Figiel and W. B. Johnson in [FiJo74] can be considered as the first descendent of Tsirelson's space. *It is a uniformly convex space that contains no copies of ℓ_p or c_0.* Then, after, say, 15 years, T. Schlumprecht constructed a more flexible variant of Tsirelson's space—presented in the Jerusalem conference in 1991 and now called the **space** S, see [Schl91]—that almost immediately created a true revolution in Banach space theory, leading to the solution of the hyperplane problem, the unconditional subbasis problem, the homogeneous problem and, above all, the creation of a HI space (as we discussed in Comments to Problem 1; see also Sect. 1.2 below). Moreover it led to results on distortable norms.

2. The first example of a **Fréchet space** (i.e., a locally convex complete metric linear space) with the property of not being isomorphic to a subspace of codimension 1 was constructed by C. Bessaga, A. Pełczyński, and S. Rolewicz in 1961 in [BePeRo61]. E. Dubinsky then proved that, in particular, *every separable Banach space has a dense subspace that is not isomorphic to its hyperplanes* [Dub71]. We refer to [PeBe79, p. 227] for more on this subject.

The fact that it took half of a century quite an effort of many world centers to do such construction for Banach spaces documents how subtle and creative the concept of Banach space is.

The **basis constant** $\mathrm{bc}(Y)$ of a Banach space Y is defined as the least upper bound of the constants L such that there is a Schauder basis $\{e_n\}_{n=1}^{\infty}$ for Y satisfying

$$\left\| \sum_{j=1}^{n} t_j e_j \right\| \le L \left\| \sum_{j=1}^{n+m} t_j e_j \right\| \text{ for all scalars } t_1, \ldots t_{n+m}, \text{ and } n, m \in \mathbb{N}.$$

If Y does not have any basis we put $\mathrm{bc}(Y) = \infty$.

If X and Y are two isomorphic Banach spaces, then the **Banach–Mazur distance** between X and Y is defined to be the infimum of $\|T\| \cdot \|T^{-1}\|$ as T ranges over all isomorphisms from X onto Y. The Banach–Mazur distance between X and Y is denoted by $d(X, Y)$.

Parts of Problem 1 are closely connected to the following more general conjecture of A. Pełczyński [Pe06].

Problem 3 (A. Pełczyński). Does there exist a constant $C \ge 1$ and a function $\varphi : [1, +\infty) \to \mathbb{R}$ with $\lim_{L \to \infty} \varphi(L) = \infty$ such that if $\dim E < \infty$ and $d(E, \ell_2^{\dim E}) \ge L$ then there is a subspace $F \subset E$ with $\mathrm{bc}(F) \le C$ and $d(F, \ell_2^{\dim F}) \ge \varphi(L)$?

Roughly speaking the conjecture says: *A finite-dimensional space which is far from a Hilbert space has a subspace which is far from a Hilbert space and has a nice basis.*

We do not know if the following problem from [LinTza77, p. 86] is still open:

Problem 4. Let $1 < p < \infty$ and let X be an infinite-dimensional Banach space that is isomorphic both to a subspace and to a quotient space of ℓ_p. Is X isomorphic to ℓ_p?

Recall that the **density of a Banach space** X is the minimal cardinality of a norm dense set in X.

Problem 5. Can the original Tsirelson's construction mentioned in Remark 6 to Problem 1, above, be adjusted to produce a reflexive Banach space X of density c with unconditional basis such that c_0 is crudely finitely representable in every infinite-dimensional subspace of X?

We defined unconditional Schauder basis at p. 1. In general, a family $\{e_\gamma\}_{\gamma \in \Gamma}$ of vectors in a Banach space X is called an **unconditional long Schauder basis** of X if for every $x \in X$ there is a unique family of real numbers $\{a_\gamma\}_{\gamma \in \Gamma}$ such that $x = \sum a_\gamma e_\gamma$ in the sense that for every $\varepsilon > 0$ there is a finite set $F \subset \Gamma$ such that $\|x - \sum_{\gamma \in F'} a_\gamma e_\gamma\| \le \varepsilon$ for every finite $F' \supset F$.

The following problem is mentioned, e.g., in [LinTza73, p. 19].

Problem 6. Assume that X is a separable Banach with unconditional Schauder basis and Y is a complemented subspace of X. Does Y have an unconditional Schauder basis? Or, at least, does such Y have a complemented subspace with an unconditional Schauder basis?

We refer to [Cass01, p. 279].

A Banach space X is called an $\mathcal{L}_p$-**space**, for $1 \leq p \leq \infty$, if there is $\lambda < \infty$ such that for every finite-dimensional subspace E of X, there is a further finite-dimensional subspace F of X with $F \supset E$ and with $d(F, \ell_p^{\dim F}) \leq \lambda$.

Problem 7. Assume that $1 < p < \infty, p \neq 2$, and that X is a separable $\mathcal{L}_p$-space. Does X have an unconditional Schauder basis?

Note that X *is then a complemented subspace of* L_p [JoLin01b, p. 57]. Thus the problem is connected with Problem 6 (use the Marcinkiewicz–Paley theorem on unconditional bases in L_p spaces for $p > 1$, see, e.g., [AlKal06, p. 130]).

This problem is in [HOS11], where more on it can be found; for example, that *any separable $\mathcal{L}_p$-space has a Schauder basis for* $1 \leq p < \infty$.

Problem 8. Let $1 < p < \infty, p \neq 2$. Assume that $L_p(\mu)$ has density $\aleph_1$ and μ is finite. Does $L_p(\mu)$ have an unconditional Schauder basis?

We took this problem from [JoSch14]. It is known that *the answer is negative if the density of the space is at least $\aleph_\omega$* [EnRo73].

Connected with Problem 8, we may ask the following:

Problem 9. Study the long Schauder bases in $L_p(\mu)$-spaces for μ finite in the sense described, e.g., in [HMVZ08, p. 132].

1.2   Hereditarily Indecomposable Spaces

An infinite-dimensional Banach space X is said to be **indecomposable** if it cannot be written as a topological direct sum of two infinite-dimensional closed subspaces of X. A Banach space X is said to be **hereditarily indecomposable** (**HI**, in short), if every closed infinite-dimensional subspace of X is indecomposable (we introduced this definition earlier, in Remark 6 to Problem 1).

The question on the existence of indecomposable spaces was first asked by J. Lindenstrauss in [Lin70]. It is not easy to find an indecomposable space that is not hereditarily indecomposable. For references see, e.g., [Mau03, p. 1251]. *There are superreflexive HI spaces* [Fer03].

From the area of HI spaces let us start by mentioning the following:

Problem 10.   Does there exist a HI space that admits a C^2-smooth bump?

Note that we mentioned in the comment preceding Problem 2 that R. Deville showed that *a HI Banach space cannot admit a C^∞-smooth bump.*

The following easily formulated problem seems to be open:

Problem 11.   Does there exist a HI space isomorphic to its dual?

We refer to [Mau03, p. 1265].

S. A. Argyros, A. D. Arvanitakis, and A. G. Tolias showed that *there is a Banach space of density c that is hereditarily indecomposable.* We refer to, e.g., [DLAT10]. Note that *every hereditarily indecomposable Banach space has to have density less than or equal to c*, see [PliYo00].

The following seems to be an open problem:

Problem 12 (S. Todorčević).   If X is a Banach space of density $> c$, does it contain an infinite-dimensional subspace with unconditional Schauder basis?

P. Koszmider posed in [Kosz10] the following open problem:

Problem 13 (P. Koszmider). Is every real HI space an extremely non-complex space?

A real Banach space X is called an **extremely non-complex space** if for every bounded linear operator T from X into X, we have

$$\|\mathrm{Id}_X + T^2\| = 1 + \|T^2\|,$$

where Id_X denotes the identity operator from X into X. We refer to [KMM09].

Problem 14 (S. A. Argyros and P. Koszmider). Is there any bound on the density of an indecomposable (not necessarily hereditarily) Banach space?

We refer to [Kosz13], where it is shown that *it is consistent that there is an indecomposable Banach space of density 2^c.*

Problem 15 (A. Pełczyński [Pe06]). Is every HI space a subspace of a HI space with a Schauder basis?

Recall that a Banach space X is called an $\mathcal{L}_\infty$-**space** if there is a real number $\lambda \geq 1$ such that for every finite-dimensional subspace E of X there is an n-dimensional subspace $F \supset E$ of X such that $\mathrm{dist}\,(F, \ell_\infty^n) \leq \lambda$. Note that *there are separable $\mathcal{L}_\infty$-spaces without isomorphic copies of c_0* [BouDel81], and thus there is a chance that the following problem has a positive solution:

Problem 16 (A. Pełczyński [Pe06]). Does there exist a HI space which is an $\mathcal{L}_\infty$-space?

J. López-Abad and S. Todorčević [LATo09] constructed a space X with Schauder basis of length ω_1 which is **saturated by copies of** c_0 (i.e., every infinite-dimensional subspace contains a space isomorphic to c_0), and such that for every decomposition of a closed subspace Y of X, say $Y = Y_0 \oplus Y_1$, in a topological direct sum, either Y_0 or Y_1 has to be separable. This is a nonseparable version of the notion of an HI space. While X is not distortable—due to the c_0-saturation—(for the definition of a distortable norm, see p. 23), it is proved that it is ω_1-**arbitrarily distortable**, meaning that for every $\lambda > 1$ there is an equivalent norm $||| \cdot |||$ on X such that for every nonseparable subspace Y of X, there exist points y_1 and y_2 in the original unit sphere of Y such that $\frac{|||y_1|||}{|||y_2|||} > \lambda$.

Problem 17. Study this new concept of a nonseparable version of HI.

1.3 Weak Hilbert Spaces

Related to the concept of weak Hilbert space (see the definition and some other information in Remark 5 to Problem 1), we formulate the following few questions:

Problem 18. Does every separable weak Hilbert Banach space have a Schauder basis?

G. Pisier proved in [Pis85] that *every weak Hilbert space has the approximation property*. A Banach space X is said to have the **approximation property** (**AP**, in short) if for every compact set K in X and for every $\varepsilon > 0$, there is a bounded, **finite-rank operator** (i.e., an operator whose range is finite-dimensional) T from X into X such that $\|Tx - x\| \leq \varepsilon$ for every $x \in K$.

It is a classical Grothendieck result (cf., e.g., [FHHMZ11, p. 701]) that X *has the approximation property if and only if for every Banach space Y, for every compact operator T from Y into X, and for every $\varepsilon > 0$, there is a bounded finite-rank operator T_1 from Y into X such that $\|T - T_1\| < \varepsilon$ (an operator $T : Y \to X$ is said to be **compact** whenever $\overline{T(B_Y)}$ is compact).

It is simple to prove that *every Banach space with a Schauder basis has the approximation property* (see, e.g., [FHHMZ11, p. 711]).

The next problem is more ambitious than Problem 18 above.

Problem 19. Let X be a separable Banach space. Is X a weak Hilbert space if and only if every subspace of X has a Schauder basis?

We mention that *there is a weak Hilbert space with no subspace isomorphic to a Hilbert space* (the 2-convexification of the modified Tsirelson space, a result of W. B. Johnson [Jo76]). On the other hand, A. Eddington showed in [Edd91] that *there is a weak Hilbert space X not isomorphic to a Hilbert space and such that X is saturated with Hilbertian spaces.* Further examples are also in [ACK00]. *Such spaces cannot admit any C^2-smooth bump function* (B. M. Makarov, see, e.g., [DeGoZi93, p. 226].

For both Problems 18 and 19 we refer to P. Casazza's paper in [Cass01, p. 277] and to G. Pisier's paper [Pis88].

———————————<>———————————

R. Komorowski proved in [Kom94] that *there is a weak Hilbert space with no unconditional basis*.

Problem 20. Is it true that every weak Hilbert space is a subspace of a weak Hilbert space with an unconditional basis?

We took this problem from [ACK00].

———————————<>———————————

We took the following problem from [ABR12]:

Problem 21 (P. Casazza). Does there exist a weak Hilbert space which is HI?

Related to this problem, we should mention that *there is a weak Hilbert space with an unconditional basis $\{e_n\}_{n=1}^{\infty}$ and having the property that, for every block subspace Y, every bounded linear operator on Y is of the form $D + S$, where D is a* **diagonal operator** *(i.e., $D(e_n) = \lambda_n e_n$ for all $n \in \mathbb{N}$, where $\{\lambda_n\}_{n=1}^{\infty}$ is some scalar sequence) and S is strictly singular* [ABR12] (a **block subspace** is the closed linear hull of a **block basic sequence**, i.e., a sequence of nonzero vectors $\{u_j\}_{j=1}^{\infty}$ of the form $u_j = \sum_{n=p_j+1}^{p_{j+1}} a_n e_n$, with $\{a_n\}_{n=1}^{\infty}$ scalars and $p_1 < p_2 < \dots$ an increasing sequence of integers). An operator $T : X \to Y$ is said to be **strictly singular** if there is no infinite-dimensional subspace Z of X such that the restriction operator $T|_Z$ is an isomorphism into Y.

Note that *every bounded linear operator T on a complex HI space has the form $\lambda Id + S$, where $\lambda \in \mathbb{C}$, Id is the identity operator and S is a strictly singular operator* (cf. [Mau03]).

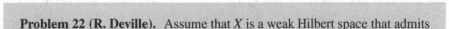

Problem 22 (R. Deville). Assume that X is a weak Hilbert space that admits a C^2-smooth bump. Is X necessarily isomorphic to a Hilbert space?

It is proved in [DeZi88] that *Problem 22 has a solution in the positive if the C^2-smooth assumption is replaced by a C^3-smooth assumption.*

Since *every weak Hilbert space is reflexive* (W. B. Johnson, see [Pis88]), the results of D. Amir and J. Lindenstrauss on weakly compactly generated spaces (see below, and see also Sect. 3.3) give that *every weak Hilbert space can be written as $Y \oplus Z$, where Y is a separable weak Hilbert space and H is isomorphic to a Hilbert space* (see [CaSh89]). A Banach space X is **weakly compactly generated—WCG**, in short—if there is a weakly compact set K in X such that X is the closed linear hull of K.

1.4   Complementability

Problem 23 (C. Bessaga, A. Pełczyński). Let K be a compact metric space and let Y be a complemented subspace of $C(K)$. Is then Y isomorphic to a $C(L)$-space for some compact metric space L?

This problem *has a positive solution for subspaces Y with nonseparable dual* (H. Rosenthal, see [Ro03])

For more information we refer to [Ro03] and [DLAT10].

The following question complements Problem 23 above:

Problem 24 (H. P. Rosenthal). Assume that K is a compact metric space and X is a complemented subspace of $C(K)$ that contains an infinite-dimensional reflexive subspace. Is X isomorphic to $C[0, 1]$?

We refer to [Ro03].

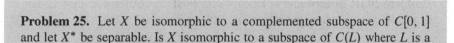

Problem 25. Let X be isomorphic to a complemented subspace of $C[0, 1]$ and let X^* be separable. Is X isomorphic to a subspace of $C(L)$ where L is a countable compact space?

We took this problem from Y. Benyamini's review of [Ro03] in Math. Reviews.

A compact K is **totally disconnected** if the only connected subsets of it are singletons. *If K is a compact metric space, then the space $C(K)$ is isomorphic to a $C(L)$-space for some compact totally disconnected space L*: Namely, to $C(D)$, where D is the Cantor set if K is uncountable (by Miljutin theorem, see, e.g., [AlKal06, p. 194]) or, if K is countable, to $C(S)$ for some ordinal segment S (due to the fact that K is then homeomorphic to S by the Mazurkiewicz–Sierpiński theorem, see, e.g., [HMVZ08, p. 73]).

It used to be a well-known open problem if a similar result holds true for nonmetrizable compacta, until recently, when P. Koszmider [Kosz04] and G. Plebanek [Ple04] constructed counterexamples (in ZFC).

In this area we chose the following two open problems from S. A. Argyros and A. D. Arvanitakis [ArgArv04]:

Problem 26 (S. A. Argyros and A. Arvanitakis). Let H be a nonseparable Hilbert space and B_H be its ball in the weak topology. Is $C(B_H)$ isomorphic to some $C(L)$-space for some totally disconnected compact space L?

Problem 27 (S. A. Argyros and A. Arvanitakis). Does there exist a weakly compact nonmetrizable convex subset K of a Banach space such that $C(K)$ is isomorphic to a $C(L)$-space for some totally disconnected compact space L?

Many open problems in this area are also contained in [Kosz10]. Here is one of them:

Problem 28 (P. Koszmider). Does there exist a $C(K)$-space that is not isomorphic to its square but it is isomorphic to its cube?

A Banach space X is called **injective** if it is complemented in every overspace. A topological space K is called **extremely disconnected** if the closure of every open subset is open.

The following is a well-known longstanding open problem:

Problem 29. Is every injective Banach space isomorphic to a $C(K)$-space for K an extremely disconnected compact space?

We refer to [Zip03]. This problem should be compared to the Sobczyk–Zippin result that *an infinite-dimensional separable Banach space X is complemented in every separable overspace if, and only if, X is isomorphic to c_0* (see, e.g., [Zip03]).

In this connection, let us note in passing that S. A. Argyros, J. F. Castillo, A. S. Granero, M. Jiménez, and J. P. Moreno proved in [ACGJM02] that *if the cardinality of Γ is less than $\aleph_\omega$, then $c_0(\Gamma)$ is complemented in WCG overspaces*. However this is not the case if the cardinality of Γ is really huge.

Problem 30. Is it true that every infinite-dimensional complemented subspace of L_1 is isomorphic either to ℓ_1 or L_1?

We refer, e.g., to [AlKal06, p. 122].

It was proven by J. Bourgain (see, e.g., [Kal06]) that *not every isomorphic copy of ℓ_1 in ℓ_1 is necessarily complemented in ℓ_1.* It was known before that *it is also not true for ℓ_p, $p > 1$, $p \neq 2$, but it is true for c_0* [FHHMZ11, p. 245]. We refer to [FHHMZ11, p. 208].

P. Koszmider asked in [Kosz06] the following question:

Problem 31 (P. Koszmider). Does there exist a separable Banach space X such that the only nontrivial decomposition of X is by a subspace isomorphic to c_0 and its complement isomorphic to X?

In the same paper he showed that *it is consistent to assume a positive answer for nonseparable spaces (of continuous functions).*

Let X be a nonseparable Banach space. Then the following are equivalent:

(i) There is a separable infinite-dimensional quotient of X, i.e., there is an infinite-dimensional separable Banach space Y and a bounded linear operator from X onto Y.

(ii) There is a separable infinite-dimensional subspace Y of X that is **quasicomplemented in** X, i.e., there is a closed subspace Z of X such that $Y \cap Z = \{0\}$ and $Y + Z$ is dense in X.

(iii) There is an infinite increasing sequence $\{X_n\}_{n=1}^{\infty}$ of distinct infinite-dimensional closed subspaces of X such that $\overline{\bigcup_{n=1}^{\infty} X_n} = X$.

For the proof we refer to [Muj77] and [HMVZ08, p. 199].

Problem 32 (S. Banach, A. Pełczyński). Does every nonseparable Banach space have the (equivalent) properties *(i)–(iii)* above?

Many spaces do have this property. For example, all WCG spaces and ℓ_∞ have it. Indeed, WCG spaces have even infinite-dimensional separable complemented subspaces (see also the remarks to Problem 33 below), and ℓ_2 is a quotient of ℓ_∞ (cf., e.g., [FHHMZ11, p. 228]). Actually, all separable subspaces of ℓ_∞ are quasicomplemented there (see, e.g., [HMVZ08, p. 200]). Moreover, all dual spaces have the equivalent properties (i)–(iii) above [ArDoKa08].

For a recent information we refer to [DLAT10]. For information related to this problem in general, we refer to [HMVZ08] and [FHHMZ11].

———————◁◎▷———————

W. B. Johnson and J. Lindenstrauss, in [JoLin74, p. 229], posed the following problem:

> **Problem 33 (W. B. Johnson and J. Lindenstrauss).** Let $X \supset Y$ be Banach spaces with Y separable. Does there exist a space Z with $X \supset Z \supset Y$, Z complemented in X and the density character of Z is less than or equal to that of the continuum c?

The answer was known to be positive with the cardinality requirement even stronger (c replaced by $\aleph_0$) for WCG spaces X (see also Sect. 3.3), due to the pioneering results of D. Amir and J. Lindenstrauss [AmLin68] that opened the gate to a large field in nonseparable spaces. Behind that all was an ingenious idea of J. Lindenstrauss to build, in nonseparable WCG spaces, by using Tychonoff compactness theorem together with Mazur's exhaustion method, a transfinite sequence of projections with smaller ranges, continuously depending on the transfinite interval topology. This is called a **projectional resolution of the identity (PRI**, in short).[1] It created a monumental amount of applications in the structure and analysis on nonseparable spaces and it took about 30 years of an effort of many schools in working on it. The state of the art, as of 2008, is described, e.g., in [DeGoZi93, Fa97], [HMVZ08, Chaps. 3 and 5] and later in [FHHMZ11, Chap. 13] (see also [Ziz03]).

Later on *Problem 33 was solved in the positive for dual spaces* by S. Heinrich and P. Mankiewicz in [HeMa82]. For a simpler proof, see [SimYo89].

Furthermore, S. P. Gulko proved that *the answer is also positive for C(K)-spaces, where K is totally disconnected. For spaces the dual ball of which are angelic in the w*-topology* the positive answer was proven by A. Plichko. A compact space is **angelic** if the closures of its subsets can be reached by convergent sequences. For these we refer to [PliYo00].

P. Koszmider proved in [Kosz05] that the negative answer to the Johnson–Lindenstrauss Problem 33 is consistent with ZFC even for spaces $C(K)$, i.e., its negation cannot be proved by the sole ZFC.

———————◁◎▷———————

---

[1]In questions on duality and smoothness, PRI substitutes in nonseparable spaces for the notion of Schauder basis and does truly miracles there.

In [Kosz05], the following open problems arose:

Problem 34 (P. Koszmider).

(a) Can a counterexample to the Johnson–Lindenstrauss Problem 33 be found in ZFC?
(b) Is the positive answer for the Johnson–Lindenstrauss Problem 33 (for $C(K)$-spaces) consistent as well?

The following problem is taken from [Kosz05] and is related to Problem 34:

Problem 35 (P. Koszmider). Is it true in ZFC that if X is a Banach space and Y is a separable subspace of X, then there is a complemented subspace Z of X of density less than or equal to 2^{ω_1} such that $Y \subset Z$?

A Banach space X has the **separable complementation property** (SCP, in short) if every separable subspace of X is contained in a separable and complemented subspace of X.

In connection with Problem 33, we note that the following seems to be open:

Problem 36 (A. Plichko). Assume that X is a Banach space having the SCP. Let $\aleph$ be an arbitrary cardinal number. Is it true that for an arbitrary subspace W of X with density $\aleph$ there is a subspace V of X with $V \supset W$ and such that V is complemented in X and has density $\aleph$?

We refer to [PliYo00].

Related to Problem 36, we note that the following is open:

Problem 37 (A. Plichko, D. Yost). Let X be a nonseparable Banach space with the Radon–Nikodým property. Does there exist a projection P on X with separable range?

A Banach space X has the **Radon–Nikodým property** (**RNP**, in short) if every bounded set in X has a slice of arbitrarily small diameter (we say then that the bounded set is **dentable**). A **slice** of a bounded set is its cut with an open halfspace.

We refer to [PliYo01]. A candidate for a counterexample cannot be a dual space [FaGo88].

Problem 38. If the space X has the SCP and Y is a complemented subspace of X, does Y have the SCP?

We took this problem from [FiPeJo11].

Problem 39. Assume that X^{**} has the SCP. Does X have the SCP?

We took this problem from [FiPeJo11].

The following few problems come from [Pe06]:

Call a Banach space X **badly locally complemented** if there is an increasing sequence $\{c_n\}_{n=1}^{\infty}$ of real numbers such that $\lim c_n = \infty$ and such that every n-dimensional subspace of X is **worse than** c_n**-complemented**. This means that any projection on any n-dimensional subspace has norm greater than or equal to c_n. In this direction we mention here that G. Pisier proved in [Pis85] that *there is an infinite-dimensional Banach space X and $\delta > 0$ such that for every finite-rank projection P on X,*

$$\|P\| \geq \delta (\operatorname{rank} P)^{\frac{1}{2}}.$$

Problem 40 (A. Pełczyński).   Does there exist a HI space that is badly locally complemented?

The following is related to Problem 40:

Problem 41 (A. Pełczyński and N. Tomczak-Jaegermann). Does there exist an increasing sequence $\{c_n\}_{n=1}^{\infty}$ with $\lim c_n = \infty$ such that for each n there is an n-dimensional Banach space F_n such that every subspace E of F_n is worse than m-complemented in F_n, where $m = \min\{\dim E, \dim F_n/E\}$?

1.5   Invariant Subspaces

Problem 42. Let T be a bounded linear operator on a complex ℓ_2 space (or just on a reflexive space). Does T have a **nontrivial invariant subspace**, i.e., does there exist a closed subspace X of ℓ_2 with $X \neq \{0\}$, $X \neq \ell_2$, and such that $TX \subset X$?

The first example of a Banach space and *an operator on it without a nontrivial invariant subspace* was constructed by P. Enflo about 40 years ago. Later on, an example of such operator was found on ℓ_1 by C. J. Read. More precisely, he constructed in [Re88] an operator on ℓ_1 such that all the nonzero vectors are hypercyclic (see Sect. 1.6) (providing even a counterexample to the **invariant subset problem**). We refer to [EnLo01], to [GroPe11], and to some notes in Sect. 1.6. The invariant subset problem for the complex ℓ_2 space (or just for a reflexive space) is, to our knowledge, open.

Problem 43. Does every adjoint operator on a Banach space have a nontrivial invariant subspace?

We refer to [EnLo01, p. 543].

⊸〇☾〇⊱⊷

1.6 Dynamics of Operators

Let X be a Banach space. A bounded linear operator $T : X \to X$ is called **hypercyclic** if there is a vector $x \in X$ whose **orbit under** T

$$\mathrm{orb}(T, x) := \{x, Tx, T^2x, \ldots\}$$

is dense in X. Every such vector x is called **hypercyclic for** T. Obviously, such a notion *only makes sense in separable spaces*. S. Rolewicz gave in [Rol69] the first example of a hypercyclic operator in a Banach space, and proved, in particular, that *there are no hypercyclic operators on $\mathbb{R}^n$ or $\mathbb{C}^n$*. In the same paper he asked whether every separable, infinite-dimensional Banach space supports a hypercyclic operator. This was solved in the positive independently by S. I. Ansari [Ans97] and L. Bernal-González [Ber99]. The theory is usually developed in the framework of locally convex Fréchet spaces, although *we can confine ourselves here to the case of infinite-dimensional separable Banach spaces*. Two very recent sources for the general setting are [BayMat09] and [GroPe11].

The connection with the invariant subspace problem (see Sect. 1.5) is the following: *An operator has no nontrivial invariant closed subset precisely when every nonzero vector is hypercyclic for the operator*. C. J. Read [Re88] showed that *such an operator does exist on ℓ_1*, leaving the problem open in Hilbert space (see Problem 42 above and the notes to it).

To recognize that a given operator $T \in B(X)$ (where X is a separable Banach space) is hypercyclic is in general not simple, and a working—a priori only sufficient—criterion, called the **Hypercyclicity Criterion**, has been widely used in the literature. It can be formulated in the following way: (Hypercyclicity Criterion): *Let $T \in B(X)$ be an operator, where X is a separable Banach space. If there are dense subsets Y and Z of X and a strictly increasing sequence $\{n_k\}_{k=1}^{\infty}$ of positive integers such that (1) for each $y \in Y$, $T^{n_k}y \to 0$, (2) for each $z \in Z$ there is a sequence $\{x_k\}_{k=1}^{\infty}$ in X with $x_k \to 0$ and $T^{n_k}x_k \to z$, then T is hypercyclic.*

The Hypercyclicity Criterion has an equivalent formulation given by J. Bès and A. Peris: *$T \in B(X)$ satisfies the Hypercyclicity Criterion if, and only if, $T \oplus T$ is hypercyclic on X^2* [BesPe99].

One of the main open problems in hypercyclicity was whether every hypercyclic operator $T : X \to X$ satisfies the Hypercyclicity Criterion. This was solved in the negative by M. De la Rosa and C. J. Reed [DRRe09] by proving that *there exists a Banach space X that supports a hypercyclic operator T such that $T \oplus T$ is not hypercyclic on X^2*.

Hypercyclicity is related to the concept of transitivity from topological dynamics (introduced by G. D. Birkhoff in the 20s). G. Godefroy and J. H. Shapiro, in [GoSha98], extended the notion of hypercyclicity to Banach spaces X not necessarily separable by calling an operator $T : X \to X$ **(topologically) transitive** if for each pair U, V of nonempty open subsets of X there is some $n \in \mathbb{N}$ with $T^n(U) \cap V \neq \emptyset$. *Every hypercyclic operator is transitive, although the converse is false in general* (see, e.g., [Bo00]). In many spaces the two concepts coincide: for example, it follows from the Baire Category Theorem that *if X is a separable Banach space, an operator $T \in B(X)$ is hypercyclic if, and only if, it is topologically transitive* [GoSha98, Theorem 1.2]). Motivated by this, F. Bayart and S. Grivaux introduced in [BayGri06] the class of frequently hypercyclic operators (again in a separable setting): An operator $T : X \to X$ is said to be **frequently hypercyclic** whenever there exists a vector $x \in X$ such that for every nonempty open subset U of X, the set $\mathbb{N}_U := \{n \in \mathbb{N} : T^n x \in U\}$ has positive **lower density**, i.e., $\liminf_{N \to \infty} \mathrm{card}\{n \leq N : n \in \mathbb{N}_U\}/N > 0$.

In [BayGri06] the following problem is formulated:

Problem 44 (F. Bayart and S. Grivaux). Is the inverse operator of an invertible frequently hypercyclic operator also frequently hypercyclic?

For hypercyclicity the answer to this problem is affirmative.

Due to the result of S. I. Ansari and L. Bernal-González mentioned above, it is natural to formulate the following question, as it appears in [BerKal01]. It is the only instance of a problem in this section dealing with nonseparable spaces.

Problem 45 (T. Bermúdez and N. J. Kalton). Is there any characterization of nonseparable Banach spaces which support a topologically transitive operator?

There is a "continuous" version of hypercyclicity in terms of a family $\{T_t : t \geq 0\}$ of operators $T_t : X \to X$ that forms a **strongly continuous semigroup of operators**, i.e., a family $\{T_t : t \geq 0\}$ that satisfies (1) $T_0 := \mathrm{Id}_X$, (2) $T_{s+t} = T_s T_t$ for all $s, t \geq 0$, and (3) the map $t \to T_t x$ is continuous on $\mathbb{R}_+$ for each $x \in X$. Then, a vector $x \in X$ is said to be **hypercyclic** for such $\{T_t : t \geq 0\}$ whenever its **orbit** $\{T_t x : t \geq 0\}$ is dense in X. From [CoPe05] we extract the following open question:

Problem 46 (A. Conejero and A. Peris). Let X be a separable Banach space. Is the orbit $\{T_t z : t \geq 0\}$ of a vector z hypercyclic under a given strong continuous semigroup $\{T_t : t \geq 0\}$ of operators a linearly independent set?

This is the case in the discrete version, i.e., when we consider instead the iteration of an operator. For details, see [CoPe05].

Let X be a Banach space. A bounded linear operator T from X into X is called **supercyclic**, if there is $x \in X$ such that the set $\{\lambda T^n x, n \geq 0, \lambda \in \mathbb{R}\}$ is dense in X. The following open question is taken from [Bay05]:

Problem 47 (F. Bayart). Let X be a separable real Banach space. Is the set of supercyclic operators from X into X spaceable in the Banach space $B(X)$ of bounded linear operators from X into X?

A subset M of a Banach space X is called **spaceable in** X if there is a closed infinite-dimensional subspace Y of X such that $M \cup \{0\} \supset Y$.

For a complex separable Banach space X the answer to Problem 47 is positive (F. Bayart).

1.7 Scalar-Plus-Compact Problem; Distortable Norms

There is an infinite-dimensional subspace X of L_p, $0 < p < 1$, such that all the bounded linear operators from X into X are scalar multiples of the identity operator Id_X (cf., e.g., [Kal03, p. 1113],) and *there is a compact convex subset of L_p, $0 < p < 1$, which has no extreme point* (cf., e.g., [Kal03, p. 1111]).

However, *there is a quasi-Banach space X* (for the definition, see Sect. 2.6) *with trivial dual space such that there is a nontrivial compact operator from X into some quasi-Banach space.* This answered a question of A. Pełczyński (see [Kal03, p. 1113]).

These two results on quasi-Banach spaces go into the same direction as the recent milestone Banach space result of S. A. Argyros and R. Haydon in [ArHay11] that *there is an infinite-dimensional separable Banach space X such that all bounded linear operators from X into X are of the form a scalar multiple of the identity operator plus a compact operator.* Note in passing that, as presented in [EnLo01] and mentioned in [ArHay11], *every operator on such X has a nontrivial invariant subspace.*

S. A. Argyros and R. Haydon result should be compared with the result of S. Shelah and J. Steprans (cf., e.g., [FHHMZ11, p. 264] and [Wa01]) that *there is a nonseparable reflexive Banach space X such that all bounded linear operators from X into X are of the form a scalar multiple of the identity operator plus an operator with separable range.*

Problem 48 (S. A. Argyros and R. Haydon). Can the example of S. A. Argyros and R. Haydon in [ArHay11] be made reflexive?

If $(X, \| \cdot \|)$ is a Banach space and $\lambda > 1$, its norm $\| \cdot \|$ (or the Banach space $(X, \| \cdot \|)$) is called λ-**distortable** if there is an equivalent norm $| \cdot |$ on X so that for all the infinite-dimensional subspaces Y of X,

$$\sup \left\{ \frac{|y_1|}{|y_2|} : y_1, y_2 \in S_{(Y, \|\cdot\|)} \right\} \geq \lambda.$$

The norm $\| \cdot \|$ (or the Banach space $(X, \| \cdot \|)$) is said to be **distortable** if it is λ-distortable for some $\lambda > 1$, and it is **arbitrarily distortable** if it is λ-distortable for all $\lambda > 1$.

R. C. James proved that *neither $(c_0, \| \cdot \|_\infty)$ nor $(\ell_1, \| \cdot \|_1)$ are distortable* (cf., e.g., [FHHMZ11, p. 275]). The existence of a distortable space was first proved by joining the work of V. D. Milman and S. Tsirelson (see [OdSc01]). Now it is known that *if an infinite-dimensional X fails to have an infinite-dimensional subspace with a distortable norm, then X must contain an isomorphic copy of c_0 or ℓ_1* (see, e.g., [BenLin00, p. 312]).

The finite-dimensional situation is different. This is here where the celebrated Dvoretzky's theorem governs: *Let f be a τ-Lipschitz function defined on the unit sphere of an n-dimensional Banach space X, and let $0 < \varepsilon < \tau/2$. Then there exists a value λ_0 and a subspace Y of X of dimension $k \geq C\varepsilon^2 \log n / |\log(\varepsilon/\tau)|\tau^2$ (where*

C is a universal constant) such that $|f(y) - \lambda_0| \leq \varepsilon$ for all $y \in Y$ with $\|y\| = 1$ (see, e.g., [BenLin00, Corollary 12.11]). For the definition of a τ-Lipschitz function, see the introduction to Chap. 5.

For the next problem see [BenLin00, p. 336].

Problem 49. Assume that X has a distortable norm. Does X contain an infinite-dimensional subspace with arbitrarily distortable norm?

The *first example of an arbitrarily distortable Banach space* was given by Th. Schlumprecht [Schl91]. Later, E. Odell and Th. Schlumprecht proved in [OdSc94] that *if* $1 < p < \infty$, *then* ℓ_p *is arbitrarily distortable* (see [OdSc01] and [BenLin00, p. 332]). N. Tomczak-Jaegermann proved that *every hereditarily indecomposable space is arbitrarily distortable*, see, e.g., [ArGoRo03, p. 1063].

It is known that *Tsirelson's space is distortable* (see, e.g., [BenLin00, p. 334]). The following question seems to be open:

Problem 50. Is Tsirelson space arbitrarily distortable?

We refer to [BenLin00, p. 335].

In the paper of J. López-Abad and S. Todorčević [LATo09] that we mentioned in Sect. 1.2, they showed that *while their space X is not distortable as it contains a copy of c_0, it is **arbitrarily nonseparably distortable*** in the sense that for every λ greater than 1, there is an equivalent norm on X such that for every nonseparable subspace Y of X there are two points on the original sphere of Y such that the ratio of their new norms is greater than λ.

We pose the following problem:

Problem 51. Study this new concept of nonseparable distortability.

1.8   Approximation Property

We repeat here that a Banach space X is said to have the **approximation property** (**AP**, in short) if for every compact set K in X and for every $\varepsilon > 0$, there is a finite-rank operator T from X into X such that $\|Tx - x\| \leq \varepsilon$ for every $x \in K$ (see the remarks after Problem 18). The following is a longstanding open problem:

> **Problem 52.** Assume that for a given Banach space X the compact operators on X can be approximated in norm by finite-rank operators (i.e., assume that for every compact operator T from X into X and for every $\varepsilon > 0$ there is a finite-rank operator T_1 from X into X such that $\|T - T_1\| < \varepsilon$). Does X have the approximation property?

We refer to [LinTza77, p. 37] and [Piet09].

The first example of a space without the approximation property—in particular, without a Schauder basis, see comments to Problem 18—was found in 1973 by P. Enflo. Later on, A. Szankowski proved that *the space of bounded operators on ℓ_2 does not have the approximation property*, and it is known now that *the spaces ℓ_p ($p \neq 2$) and c_0 all have closed subspaces without the approximation property*. Nowadays, this concept is better and better understood, but, as Problem 52 shows, still there is a need for further research in this direction. We refer to [FHHMZ11, p. 711] and [Cass01].

> **Problem 53.** It is apparently unknown if the nonseparable Banach space H^∞ of **bounded analytic functions** (i.e., functions admitting locally a power series expansion) on the open unit disk with the supremum norm has the approximation property.

We refer to [Cass01, p. 285].

Let G be the usual **Circle Group**, $M(G)$ be the set of all complex valued measures on G with bounded variation. Furthermore let A be the set of the exponents $\{e^{i2^k t} : k = 0, 1, \ldots\}$ and

$$M_{A\perp}(G) = \left\{ \mu \in M(G) : \int_G \overline{\gamma(x)}\mu(dx) = 0 \text{ for } \gamma \in A \right\}.$$

Then $M_{A^\perp}(G)$, under the norm of total variation of measure, is a dual Banach space.

Problem 54 (A. Pełczyński). In this notation, does $M_{A^\perp}(G)$ have the approximation property?

We refer to [Pe06].

———※◎※———

Problem 55 (W. B. Johnson and A. Szankowski). Assume that all subspaces of a given Banach space X have the approximation property. Is X necessarily reflexive?

We took this problem from [JoSz12].

———※◎※———

Let $1 \le \lambda < \infty$. A Banach pace X is said to have the λ-**approximation property** if for every $\varepsilon > 0$ and every compact set K in X, there is a finite-rank operator T from X into X so that $\|T\| \le \lambda$ and $\|Tx - x\| \le \varepsilon$ for every $x \in K$. A Banach space X is said to have the **bounded approximation property** (**BAP**, in short) if it has the λ-approximation property for some λ. A Banach space is said to have the **metric approximation property** (**MAP**, in short) if it has the 1-approximation property.

Problem 56. Does there exist a subspace of c_0 that has the approximation property and such that it does not have the bounded approximation property?

We refer to [Go01a].

———※◎※———

Problem 57. Let X have the bounded approximation property. Does there exist an equivalent norm $||| \cdot |||$ on X such that $(X, ||| \cdot |||)$ has the metric approximation property?

We refer to [LinTza77, p. 42] and [Cass01, p. 290].

Problem 58. Does ℓ_1 or ℓ_∞ have the metric approximation property for every equivalent norm on the space?

We refer to [Cass01, p. 290]. *The space c_0 does have this property*, see [Cass01, p. 290].

The following is a longstanding open problem:

Problem 59. Let X be a nonseparable dual Banach space with the approximation property. Must X have the metric approximation property?

For separable spaces the answer is affirmative (A. Grothendieck, see, e.g., [Cass01, p. 289]). More generally, *it is also positive if the dual space X has the RNP* (Grothendieck, see, e.g., [FHHMZ11, p. 720]).

If n is a natural number, then $BV(\mathbb{R}^n)$ denotes the space of all real-valued functions with bounded variation on $\mathbb{R}^n$, i.e., the functions u in $L^1(\mathbb{R}^n)$ whose distributional gradient Du is a bounded vector-valued Radon measure. Then $\|u\|_{BV} := \|u\|_1 + \|Du\|$, where $\|Du\|$ is the total variation of the measure Du.

Problem 60 (T. Figiel, W. B. Johnson and A. Pełczyński). Does $BV(\mathbb{R}^n)$ have the metric approximation property for $n = 2, 3, \ldots$?

We refer to [Pe06]. We remark that G. Alberti, M. Csörnyei, A. Pełczyński, and D. Preiss proved in [ACPP05] that this space has the bounded approximation property.

We say that a Banach space X has the **strong approximation property** if the following holds: For every separable reflexive Banach space Z and for every compact operator T from X into Z, there is a **bounded net**[2] of finite-rank operators T_α from X into Z such that $\|T_\alpha x - Tx\| \to 0$ for every $x \in X$.

Related to Problem 59 is the following problem from [Oja08]:

Problem 61. Does the strong approximation property imply the bounded approximation property?

A Banach space X has the **compact approximation property** (**KAP**, in short) if for every $\varepsilon > 0$ and every compact set K in X there is a compact operator T so that $\|x - Tx\| \le \varepsilon$ for all $x \in K$. Obviously, the approximation property implies the compact approximation property.

There is a Banach space X that has the compact approximation property and not the approximation property, see, e.g., [Cass01, p. 309]. *There are Banach spaces without the compact approximation property*, see, e.g., [Cass01, p. 308].

Problem 62. If X^* has the compact approximation property, must X have the compact approximation property?

We refer to [Cass01, p. 309].

A Banach space X is said to have the π_λ-**property** if there is a net $\{S_\alpha\}$ of finite-rank projections on X converging strongly to the identity operator on X (i.e., $\|S_\alpha x - x\| \to 0$ for every $x \in X$), and $\limsup_\alpha \|S_\alpha\| \le \lambda$. If a space X has the property π_λ for some λ it is said to have **the π-property**.

---

[2]For the definition, see, e.g., [FHHMZ11, p. 735].

Note that every space with a Schauder basis has the π-property, but the converse is not true, see, e.g., [Cass01, p. 301].

A sequence $\{X_n\}_{n=1}^{\infty}$ of finite-dimensional subspaces of a Banach space X is called a **finite-dimensional decomposition of** X (**FDD**, in short) if every $x \in X$ can be uniquely written as $x = \sum_{n=1}^{\infty} x_n$, $x_n \in X_n$, $n \in \mathbb{N}$. If $P_n(x) := \sum_{i=1}^{n} x_i$, then P_n is called the n-**th projection associated to the FDD**. It is clear that a space with FDD has the π-property. For information on FDD's see, e.g., [LinTza77, Sect. 1.g].

Problem 63. Does every separable Banach space X with the π-property have a finite-dimensional decomposition?

We refer to [Cass01, p. 299].

The following is a nonlinear version of the approximation property: a separable Banach pace is said to be **approximable** if there is a sequence $\{\psi_n\}_{n=1}^{\infty}$ of uniformly equicontinuous maps from X into X with relatively compact ranges, such that $\lim_n \|x - \psi_n(x)\| = 0$ for every $x \in X$. We refer to [Kal12]. It was shown in this paper that *if X^* is separable, then X is approximable*, and it was shown in [GoKal03] that *if we change uniform equicontinuity to uniform equiLipschitz, then we get exactly the bounded approximation property* (a family $\{\psi_i\}$ of functions is **equiLipschitz** if any of them is Lipschitz, and there is a common Lipschitz constant for all of them). We state the following problem from [Kal12]:

Problem 64 (N. J. Kalton). Is every (separable) Banach space approximable?

We pose the following problem:

Problem 65. Study further the property of being approximable.

Compare with, e.g., [FHHMZ11, Chap. 16].

1.9   Extension of Operators

Let X be a Banach space, let E be a subspace of X, and let $\lambda \geq 1$. We say that
the pair (E, X) has the λ-$C(K)$-**Extension Property** (λ-$C(K)$-EP, for short) if for
every compact space K, any bounded linear operator $T : E \to C(K)$ admits an
extension $\hat{T} : X \to C(K)$ with $\|\hat{T}\| \leq \lambda \|T\|$. We say that the pair (E, X) has the
$C(K)$-**Extension Property** ($C(K)$-EP, for short) if it has the λ-$C(K)$-EP for some
$\lambda \geq 1$.

Problem 66 (N. J. Kalton). Does there exist a separable superreflexive
space X such that $(X, C[0, 1])$ has the $C(K)$-EP?

We refer to [Kal06, p. 167].
It was proved by W. B. Johnson and M. Zippin in [JoZi89] that *if E is a subspace
of $X = c_0(\Gamma)$, where Γ is an arbitrary set, then the pair (E, X) has the $(1 + \varepsilon)$-
$C(K)$-EP for every $\varepsilon > 0$, but not for $\varepsilon = 0$.* We refer to [Zip03].

Problem 67. If X is a reflexive space with Gâteaux differentiable norm, E is
a subspace of X such that (E, X) has the $(1 + \varepsilon)$-$C(K)$-EP for every $\varepsilon > 0$,
does (E, X) have the 1-$C(K)$-EP?

This is *true for uniformly smooth Banach spaces X*. We refer to [Zip03, p. 1732].

The following problem relates to the results of G. J. Minty, A. D. Kirsbraun, and
S. Banach, see [Min70] and [BenLin00, pp. 18 and 21].

Problem 68 (N. J. Kalton). If $n \in \mathbb{N}$ and $\frac{1}{2} < \alpha < 1$, what is the least
constant c_α so that whenever M is a metric space, E a subset of M and $\varphi :
E \to \ell_2^n$ is a map satisfying:

$$\|\varphi(x) - \varphi(y)\| \leq d(x, y)^\alpha, \quad x, y \in E,$$

(continued)

Problem 68 (continued)
then there is an extension $\psi : M \to \ell_2^n$ with

$$\|\psi(x) - \psi(y)\| \leq c_\alpha d(x,y)^\alpha, \quad x, y \in M.$$

Is it true that $c_\alpha \leq Cn^{\alpha - \frac{1}{2}}$, where C is some constant?

We took the problem from N. J. Kalton in [Kal04] and from A. Naor and Y. Rabani in [NaRa]. G. J. Minty showed in [Min70] that *if $\alpha \leq \frac{1}{2}$, then $c_\alpha = 1$ works.*

1.10 Retractions

If A is a subset of a metric space X, then a **retraction** from X onto A is a map from X onto A such that f is the identity on A. If such a mapping exists, we say that A is a **retract** of X. If the mapping f is continuous (uniformly continuous, Lipschitz, etc.) we say that f is a **continuous (uniformly continuous, Lipschitz, etc.**, respectively) **retraction**, and that A is a **continuous (uniformly continuous, Lipschitz, etc.**, respectively) **retract** of X.

Problem 69. Is the ball of a separable or even WCG Banach space X a uniformly continuous retract of the ball of its second conjugate X^{**}?

In general spaces there is a counterexample due to N. J. Kalton. We refer to [Kal11] and [GoLaZi14].

Problem 70 (G. Godefroy and N. Ozawa). Let X be a separable Banach space. Does there exist a compact convex subset of X containing 0 that **generates** X (i.e., the closed linear hull of K equals X) and is a Lipschitz retract of X?

The answer to this problem is positive if X has an unconditional basis, see [GoOz14].

1.11 Nuclear Operators

A bounded linear operator T from a Banach space X into a Banach space Y is called **nuclear** if there is a sequence $\{x_n^*\}_{n=1}^{\infty}$ in X^* and a sequence $\{y_n\}_{n=1}^{\infty}$ in Y such that

$$T(x) = \sum_{n=1}^{\infty} x_n^*(x) y_n \quad \text{for all } x \in X, \text{ and} \quad \sum_{n=1}^{\infty} \|x_n^*\| \cdot \|y_n\| < \infty.$$

The set of all nuclear operators from X into X is usually denoted by $\mathcal{N}(X)$. *Every nuclear operator is compact.*

If $B(X)$ denotes the set of all bounded linear operators on X, $\mathbb{C}$ the set of complex numbers, and Id_X is the identity operator on X, we formulate the following:

Problem 71 (G. Pisier). Does there exist a complex Banach space X such that

$$B(X) = \bigcup_{\lambda \in \mathbb{C}} \{\lambda \mathrm{Id}_X + \mathcal{N}(X)\}?$$

We refer to [Piet09]. We note that a positive answer to this problem would solve Problem 52 in the negative, see again [Piet09]. Compare also with the result of S. A. Argyros and R. Haydon mentioned in Remarks to Problem 48.

Problem 72. Does there exist an infinite-dimensional Banach space X such that $\mathcal{N}(X) = K(X)$, where $K(X)$ denotes the space of all compact operators on X?

We refer to [Piet09].

A complex Banach space X having the approximation property is called a **Lidskii space** if every nuclear operator T on X with absolutely summable eigenvalues $\lambda_i(T)$ counted according to their multiplicity, satisfies the **trace formula**, i.e., $\sum x_i^*(x_i) = \sum \lambda_i(T)$, where $\{x_i\}_{i=1}^\infty$ and $\{x_i^*\}_{i=1}^\infty$ are the sequences in X and X^*, respectively, associated to the above definition of a nuclear operator. For the correctness of this definition see, e.g., [LinTza77, Theorem 1e.4].

Problem 73 (W. B. Johnson and A. Szankowski). Is X a Lidskii space if, and only if, every subspace of X has the approximation property?

We took this problem from [JoSz14]. We refer also to [Pis88].

If X is a Banach space, put $d_n(X) := \sup\{d(E, \ell_2^n) : E \subset X, \dim E = n\}$, where $d(E, \ell_2^n)$ denotes the Banach–Mazur distance (for the definition, see Problem 3). Then we can formulate:

Problem 74 (W. B. Johnson and A. Szankowski). Does every subspace of X have the approximation property if $d_n(X) = o(\ln n)$ for $n \to \infty$?

We took this problem from [JoSz12].

Problem 75 (W. B. Johnson and A. Szankowski). Assume that, for separable X, $d_n(X)$ grows up to infinity sufficiently slowly. Does X admits a finite-dimensional decomposition?

We took this problem from [JoSz12].

1.12   The Entropy Number of an Operator

Let X and Y be Banach spaces, and let $T : X \to Y$ be a bounded linear operator. For $n \in \mathbb{N}$, the **n-th entropy number** $e_n(T)$ of T is defined as the infimum of all $\varepsilon > 0$ such that the image of the unit ball B_X under T can be covered by 2^{n-1} translates of the ε-ball of Y, i.e.,

$$T(B_X) \subset \bigcup_{i=1}^{2^{n-1}} (y_i + \varepsilon B_Y), \quad \text{for some } y_1, \ldots, y_{2^{n-1}} \text{ in } Y.$$

For $0 < p < \infty$, put

$$\mathcal{L}_p^{ent} := \left\{ T \in B(X) : \sum_{n=1}^{\infty} (e_n(T))^p < \infty \right\}$$

Problem 76. If X is a Banach space, is $T \in \mathcal{L}_p^{ent}$ if and only if $T^* \in \mathcal{L}_p^{ent}$?

We refer to [Piet09].

1.13   The Daugavet Property

We say that a Banach space X has the **Daugavet property** if

$$\|\mathrm{Id}_X + T\| = 1 + \|T\| \tag{1.1}$$

for every rank-1 operator $T : X \to X$, where Id_X denotes the identity operator from X onto X. I. K. Daugavet proved in 1963 that $C[0, 1]$ *satisfies this equation for all compact operators T from $C[0, 1]$ into itself*. Since then the study of this property and its application has been a branch in the Banach space area.

We just remark here that *no space with this property has the RNP*, and *every space with the Daugavet property satisfies equation* (1.1) *for every weakly compact operator T*.

We refer to [Wer01] and mention a problem from there:

Problem 77. If X has the Daugavet property, must X contain an isomorphic copy of ℓ_2?

1.14 The Numerical Index

The next problem is meant to touch on the area of numerical index of Banach spaces. In the following problem, X will be a real Banach space. We need a few definitions. First, if X is a real Banach space and $T : X \rightarrow X$ is a bounded operator, then the **numerical range of** T is defined by

$$V(T) := \{x^*(Tx) : x^* \in S_{X^*},\ x \in S_X,\ x^*(x) = 1\},$$

and the **numerical radius of** T is defined by

$$v(T) := \sup\{|\lambda| : \lambda \in V(T)\}.$$

Finally, the **numerical index of the space** X is defined by

$$n(X) := \inf\{v(T) : T \in B(X),\ \|T\| = 1\},$$

where $B(X)$ denotes the set of all bounded operators from X into X.

By using the rotations, it follows that if X happens to be a real Hilbert space, then there is a norm-one operator on X such that its numerical range consists of the zero only, so $n(X) = 0$ then.

However, the following seems to be open:

Problem 78 (D. Li, C. Finet). In the notation above, is $n(\ell_p) > 0$ for all $p \neq 2$?

We followed here [KMP00].

Chapter 2
Basic Linear Geometry

2.1 Chebyshev Sets

A subset K of a Banach space X is said to be a **Chebyshev set** if every point in X has a unique nearest point in K. In such a case, the mapping that to $x \in X$ associates the point in K at minimum distance is called the **metric projection**.

V. Klee showed that *if the cardinality of Γ is, for example, the continuum c, then $\ell_1(\Gamma)$ can be covered by pairwise disjoint shifts of its closed unit ball* (we refer, e.g., to [FLP01]).

Note that the centers of these shifts in Klee's result clearly form a Chebyshev set that is not convex.

It is simple to prove that *every closed convex subset of a strictly convex reflexive Banach space is Chebyshev*. The norm $\|\cdot\|$ of a Banach space X is said to be **strictly convex** (or **rotund**) if its unit sphere does not contain nontrivial segments.

The existence of a nonconvex Chebyshev set in ℓ_2 is a longstanding open problem, which is equivalent with the statement that there is a nonsingleton set S in ℓ_2 such that each point of ℓ_2 has a unique **farthest point** in S (see, e.g., [FLP01]).

Thus we formulate

Problem 79. Is every Chebyshev set in ℓ_2 convex?

We refer to [FLP01]. It is known that *in a smooth finite-dimensional normed space X, every Chebyshev subset K is convex, and the metric projection is continuous on $X \setminus K$*. A result of Vlasov [Vla70] is that *in a Banach space X with a strictly convex dual norm, every Chebyshev subset with a continuous metric projection is convex*. Theorem 3.5.9 in [BorVan10] gives the result of V. Klee that *every weakly closed Chebyshev set in ℓ_2 is convex*.

© Springer International Publishing Switzerland 2016   37
A.J. Guirao et al., *Open Problems in the Geometry and Analysis of Banach Spaces*,
DOI 10.1007/978-3-319-33572-8_2

We recommend the recent paper [FleMoo15]. In this paper, *a nonconvex Chebyshev set is constructed in a noncomplete inner product space*.

━━━◦◎◦━━━

A subset K of a Banach space X is called **antiproximinal**, if no point in $X \setminus K$ has a nearest point in K. The first such closed bounded convex set K with nonempty interior was constructed in c_0 by M. Edelstein and A. C. Thompson in [EdTh72]. It was also showed there that this means that *there is no nonzero functional on c_0 that would be supporting for K and B_X as well* (a nonzero functional f on a Banach space X is said to be a **support functional** of a bounded subset D of X if there exists a point $x_0 \in D$ such that $f(x_0) = \sup\{f(x) : x \in D\}$; the point x_0 is said to be a **support point** of D; the set of all support functionals of D is denoted by $S(D)$). Thus such situation cannot happen in any space with the RNP, since there are supporting functionals forming residual sets in the dual space (cf., e.g., [FLP01, p. 634]).

Now, if $\| \cdot \|_1$ with closed unit ball B^1 and $\| \cdot \|_2$ with closed unit ball B^2 are two equivalent norms on a Banach space X, the norms $\| \cdot \|_1$ and $\| \cdot \|_2$ are called **companion norms** if B^1 is an antiproximinal set in $(X, \| \cdot \|_2)$, which obviously gives also that B^2 is an antiproximinal set in $(X, \| \cdot \|_1)$ as well.

We refer to [BJM02], where the following open problem was posted:

> **Problem 80 (J. M. Borwein, M. Jiménez-Sevilla, and J. P. Moreno).** Let X be a Banach space. Is the set of equivalent norms on X that admit a companion norm first Baire category in the space $N(X)$, where $N(X)$ denotes the metric space of all equivalent norms on X with the metric of uniform convergence on the unit ball of X?

2.2   Tilings

A **tiling** of a Banach space X is a representation of X as a union of sets with pairwise disjoint interiors. We will restrict here to tilings with bounded convex closed sets having a nonempty interior. It follows directly that the space c_0 can be tiled with shifts of the $\frac{1}{2}$-multiple of the closed unit ball to points with integer coordinates (see [FoLin98]). In this direction the following is an open problem:

> **Problem 81 (V. Fonf and J. Lindenstrauss).** Does there exist a reflexive Banach space that can be tiled by shifts of a single closed convex bounded set with nonempty interior?

We refer to [FoLin98] and to [FLP01] for more on this problem.

We can also ask

Problem 82. For which separable reflexive Banach spaces does there exist a tiling such that there are positive numbers r and R such that every tile in this tiling has two properties: it contains a ball with radius r and is contained in a ball of radius R?

We note that D. Preiss recently proved in [Pr10] that ℓ_2 *is such a space.*

2.3 Lifting Quotient Maps

Let T be a bounded linear operator from a Banach space X onto a Banach space Y, with kernel $Z \subset X$. Note that Y can be identified with X/Z by the Banach Open Mapping principle (and so T is identified to the quotient mapping $X \to X/Z$).

We search for a map f (in general nonlinear) from Y into X such that $T(f(y)) = y$ for every $y \in Y$. This mapping is called a **lifting** of the mapping T. The classical Bartle–Graves theorem (see, e.g., [FHHMZ11, Corollary 7.56]) says that *there always exists such a continuous lifting* f (we may assume $f(0) = 0$). Then the map h from X onto $Z \oplus Y$ defined by $h(x) = (x - f(Tx), Tx)$ is a nonlinear homeomorphism. Note that this map is a Lipschitz homeomorphism if f is a Lipschitz map.

In general there is no uniformly continuous lifting of such T (see, e.g., [BenLin00, p. 24]). Note that if T has a bounded linear lifting f then the mapping $P = f \circ T$ from X into X is a bounded linear projection from X onto $f(Y)$, and $f(Y)$ is linearly isomorphic to Y.

I. Aharoni and J. Lindenstrauss proved in [AhaLi78] (see, e.g., [FHHMZ11, p. 640]) that *there is a nonseparable Banach space* X, *a Banach space* Y *isomorphic to* $c_0(\Gamma)$ *for* Γ *uncountable, and a quotient map* T *from* X *onto* Y, *such that* (1) *the kernel of* T *is isomorphic to* c_0, (2) T *has a Lipschitz lifting, and* (3) X *is not linearly isomorphic to* $c_0(I)$ *for any uncountable* I. Since $c_0(\Gamma) \oplus c_0$ is isomorphic to such $c_0(I)$, we get an example of *two nonseparable spaces that are Lipschitz homeomorphic and not linearly isomorphic.* We will come to these questions later on again.

Due to the power of the concept of differentiability, the situation in separable spaces is completely different. This is due to the following result of G. Godefroy and N. J. Kalton in [GoKal03] (see also [Kal08]): *If Y is separable and Q is a bounded linear operator from X onto Y that admits a Lipschitz lifting f, then Q admits a bounded linear lifting T with* $\|T\| \leq \mathrm{Lip}(f)$, where $\mathrm{Lip}(f)$ is the Lipschitz constant of f, i.e., $\sup\{\|f(x_1) - f(x_2)\|/\|x_1 - x_2\| : x_1, x_2 \in X, x_1 \neq x_2\}$. This has the following corollary [GoKal03] (see also [Kal08]): *If a separable Banach space X is nonlinearly isometric to a subset of a Banach space Y then X is linearly isometric to a **contractively complemented subspace** of Y*, i.e., by a projection of norm 1. Indeed, denote the nonlinear isometry of X into Y by f and assume without loss of generality that the closed linear hull of $f(X)$ is Y. Assume that $f(0) = 0$. Then by a result of Figiel (see, e.g., [BenLin00, p. 342]), there is a bounded linear operator T from Y onto X such that $T(f(x)) = x$ for all $x \in X$ and $\|T\| = 1$. So, T is a quotient map from Y onto X that has a nonlinear lifting that is in fact an isometry, and we can use the Godefroy–Kalton result mentioned above.

The following is a slightly modified open question by G. Godefroy in [Go10]:

Problem 83 (G. Godefroy). Let X and Y be separable Banach spaces. Does there exist a sequence of contractively complemented subspaces of Y, each of them linearly isometric to X, that would guarantee the existence of a nonlinear isometry f from X into Y with the closed linear hull of $f(X)$ being Y?

2.4   Isometries

The following is an old classical open problem:

Problem 84. Assume that X is an infinite-dimensional separable Banach space such that for any pair of points x and y in the unit sphere of X there is a linear isometry T from X onto X such that $Tx = y$. Is X linearly isometric to a Hilbert space?

It does hold for finite-dimensional spaces. However, it does not hold true for nonseparable spaces. We refer to [PeBe79, p. 255].

We do not know if the following is still open:

Problem 85. Assume that an infinite-dimensional Banach space X is such that both X and X^* are linearly isometric to subspaces of L_1. Is X linearly isometric to a Hilbert space?

The isomorphic version works, see [BePe75, p. 257].

———————⟨◎⟩———————

The following problem is related to the classical Mazur–Ulam theorem, stating that *if X and Y are Banach spaces, and $T : X \to Y$ is a surjective isometry, then T is affine* (cf., e.g., [FHHMZ11, p. 548]).

Problem 86 (D. Tingley [Tin87]). Let X and Y be real Banach spaces. Suppose that $T_0 : S_X \to S_Y$ is a surjective isometry. Does T_0 have a linear isometric extension $T : X \to Y$?

P. Mankiewicz [Man72] proved that *if $U \subset X$ and $V \subset Y$ are open and connected, then every isometry from U onto V can be extended to an affine isometry from X onto Y.* Tingley's problem *has an affirmative answer for finite-dimensional polyhedral Banach spaces* [KaMar12]. For references and further information, see [Tana14].

———————⟨◎⟩———————

Surprisingly, the next two-dimensional problem, a particular case of Problem 86, remains apparently open.

Problem 87. Does Problem 86 have a positive answer for $Y = X$ and $\dim (X) = 2$?

We mention [Tana14], where a history of these problems and a list of references are included.

We finish this section by mentioning in passing that K. Jarosz showed that *every real Banach space X can be equivalently renormed so that all linear isometries of X onto itself are $\pm\, Id_X$* [Ja88]. See, e.g., [HMVZ08, p. 297].

2.5   Quasitransitive Norms

A norm of a Banach space X is called **quasitransitive** if for every $x, x' \in S_X$ and every $\varepsilon > 0$ there is a linear isometry S from X onto X such that $\|S(x) - x'\| < \varepsilon$. It follows that *the canonical norm on L_p is quasitransitive* (see, e.g., [DeGoZi93, p. 163]), while S. J. Dilworth and B. Randrianantoanina proved in [DilRan15] that ℓ_p *for $1 < p < \infty$, $p \neq 2$, does not have any equivalent quasitransitive norm.*

> **Problem 88.** Assume that X is an infinite-dimensional separable Banach space such that any subspace of X admits an equivalent quasitransitive norm. Is X necessarily isomorphic to a Hilbert space?

We refer to [DilRan15].

2.6   Quasi-Banach Spaces

A map $x \to \|x\|$ from a vector space X into $[0, \infty)$ is called a **quasi-norm** if

1. $\|x\| > 0$ if $x \neq 0$,
2. $\|\alpha x\| = |\alpha| \|x\|$ for a scalar α and for $x \in X$,
3. There is a constant $C \geq 1$ such that $\|x_1 + x_2\| \leq C(\|x_1\| + \|x_2\|)$ for all $x_1, x_2 \in X$.

The constant C is called the **modulus of concavity of the quasi-norm** $\|\cdot\|$.

For $0 < p \leq 1$, we call $\|\cdot\|$ a *p*-**norm** if, in addition of being a quasi-norm, it is *p*-**subadditive**, that is

4. $\|x_1 + x_2\|^p \leq \|x_1\|^p + \|x_2\|^p$ for all $x_1, x_2 \in X$.

It is known that *every quasi-norm is equivalent to a p-norm for some $0 < p \leq 1$.*

If $\|\cdot\|$ is a *p*-norm, then the function $\|x - y\|^p$ is a metric on X, and X equipped with $\|\cdot\|$ is called a **quasi-Banach space** if it is complete in this metric.

An example of a quasi-Banach space is the L_p space of all *p*-power integrable functions with the quasi-norm $\|f\| := \left(\int |f|^p\right)^{\frac{1}{p}}$ for $0 < p < 1$. Its modulus of concavity is $C := 2^{\frac{1}{p}-1}$, and $\|\cdot\|$ is a *p*-norm.

Quasi-Banach spaces suffer from the lack of the Hahn–Banach theorem. For example, *if $0 < p < 1$, then $L_p^* = \{0\}$* (M. M. Day). This leads for example to the following open problem:

Problem 89. Does every infinite-dimensional quasi-Banach space have a proper infinite-dimensional closed subspace?

By "proper" we mean here "different from the whole space."
N. J. Kalton constructed in [Kal65] *a quasi-Banach space X that contains a vector x ≠ 0 such that every closed infinite-dimensional subspace of X contains x.*

Problem 90 (Kwapien). Assume that X is a Banach space that is isomorphic to a subspace of L_p, $0 < p < 1$. Is X necessarily isomorphic to a subspace of L_1?

We refer to [BenLin00, p. 195]. *The answer is positive if X is reflexive,* and *there is a 3-dimensional Banach space that is not isometric to any subspace of L_1* (A. Koldobsky). For both results see [BenLin00, p. 195]. In this direction note that *any 2-dimensional Banach space is isometric to a subspace of L_1* (J. Lindenstrauss, see, e.g., [FHHMZ11, p. 266]).

2.7 Banach–Mazur Distance

If X and Y are two isomorphic Banach spaces, recall that $d(X, Y)$ denotes the **Banach–Mazur distance from X to Y**, i.e., the infimum over all isomorphisms T from X onto Y of the set $\{\|T\| \cdot \|T^{-1}\|\}$. If X is a Banach space, let

$$D(X) := \sup\{d(Y, Z) : Y, Z \text{ are isomorphic to } X\}.$$

It is known that *there is a constant $C > 0$ such that if the dimension of X is n, then $Cn \leq D(X) \leq n$* (see, e.g., [JoOd05]).

Problem 91 (W. B. Johnson and E. Odell). Let X be a nonseparable Banach space. In the notation above, is $D(X)$ infinite?

We note that W. B. Johnson and E. Odell showed in [JoOd05] that *the answer to Problem* 91 *is positive for separable infinite-dimensional spaces*. It is known that $D(X)$ *is also infinite for superreflexive spaces* (V. I. Gurarii, see [JoOd05]). G. Godefroy showed in [Go10a] that $D(X)$ *is infinite for all nonseparable subspaces of* ℓ_∞ *under Martin's MM axiom* (see Chap. 3), *and that in ZFC, it is infinite for all nonseparable so-called* **representable** *subspaces of* ℓ_∞.

Problem 92. Is Godefroy's result mentioned in comments to Problem 91 valid in ZFC for all nonseparable subspaces of ℓ_∞?

This question is raised in [Go10a].

Using projectional resolutions of the identity (cf., e.g., [DeGoZi93] and [HMVZ08]) the following special case of Problem 91 may be perhaps easier to decide on:

Problem 93. If X is a nonseparable WCG space, is $D(X)$ infinite?

Problem 94. Let X be a Banach space such that X^{**}/X is infinite-dimensional and let $\lambda \geq 1$. Does there exist an equivalent norm $\| \cdot \|_\lambda$ on X such that the Banach–Mazur distance of $(X, \| \cdot \|_\lambda)$ to any isometric dual space is $\geq \lambda$?

We refer to [Sin78] and to [GoSa88].

A topological space is **hereditarily separable** if every subspace of it is separable. A compact space is called **scattered** if every closed subspace L of it has an isolated point in L.

Under a consistent axiom, there is a scattered nonmetrizable compact space K such that $C(K)^*$ is hereditarily separable in its w^*-topology, and $C(K)$ is hereditarily Lindelöf in its weak topology.

We will call such compact the **Kunen compact**, see, e.g., [HMVZ08, p. 151].

Since $C(K)$ has a special renorming property (see [Go10a]), the following problem is of a special interest:

Problem 95. If K denotes the Kunen compact, is $D(C(K))$ infinite?

Problem 96 (E. Odell). Is it true that for every infinite-dimensional separable Banach space X there exist two spaces Y and Z isomorphic to X such that the Banach–Mazur distance $d(Y, Z) = 1$ and Y is not isometric to Z?

For the space $X = c_0$ the solution in the positive (due to C. Bessaga and A. Pełczyński) is in [FHHMZ11, p. 276].

Problem 97 (A. Pełczyński). Determine the asymptotic behavior of the Banach–Mazur distance function from the cube, i.e., of the function $R^n_\infty :=$ $\max\{\text{dist}\,(X, \ell^n_\infty) : \dim (X) = n\}$ as $n \to \infty$.

It is known that for some absolute constants c and C, $c\sqrt{n}\ln n \le R^n_\infty \le Cn^{\frac{5}{6}}$. We refer to [To97].

Following [JoOd05], call a Banach space X K-**elastic** if there is a constant $K > 0$ such that for every equivalent norm $\|\|\cdot\|\|$ on X, $(X, \|\|\cdot\|\|)$ is K-linearly isomorphic to a subspace of X. For example, *the space $C[0, 1]$ is 1-elastic* (see, e.g., [FHHMZ11, p. 240]).

Problem 98 (W. B. Johnson and E. Odell). Assume that an infinite-dimensional Banach space X is K-elastic for some $K > 0$. Is it true that every Banach space of density equal to that of X isomorphically embeds into X?

W. B. Johnson and E. Odell showed in [JoOd05] that c_0 *isomorphically embeds into X if X is separable and K-elastic for some $K > 0$.*

D. Alspach and B. Sari recently showed that *Problem* 98 *has a positive solution for* dens $X = \aleph_0$ [AlsSa16].

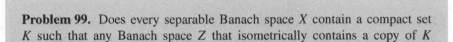

Problem 99. Does every separable Banach space X contain a compact set K such that any Banach space Z that isometrically contains a copy of K necessarily contains a linear isometric copy of X?

We refer to [DuLan08].

2.8 Rotund Renormings of Banach Spaces

The norm $\| \cdot \|$ of a Banach space X is said to be **locally uniformly rotund** (**LUR**, in short) if the following holds: Given an arbitrary $x \in S_X$ and a sequence $\{x_n\}_{n=1}^{\infty}$ in S_X such that $\|x + x_n\| \to 2$, then $\|x - x_n\| \to 0$. Clearly, an LUR norm is strictly convex. Note that if the dual norm of a Banach space $(X, \| \cdot \|)$ is LUR, then $\| \cdot \|$ is Fréchet differentiable (see, e.g., [FHHMZ11, Corollary 7.25]).

The notion of LUR was introduced in the thesis of A. Lovaglia [Lov55] under the supervision of R. C. James. Soon it became extremely useful in many areas of Banach spaces. M. I. Kadets *renormed every separable space by such a norm* (see, e.g., [FHHMZ11, p. 383]) and used it to solve a 40-year-old M. Fréchet open problem in the positive, namely whether *all infinite-dimensional separable Banach spaces are mutually homeomorphic* (see, e.g., [FHHMZ11, p. 543]). J. Lindenstrauss used it in a substantial strengthening of the Krein–Milman theorem [Lin63]. J. Lindenstrauss [Lin63] and independently E. Asplund [As68] used it to show *the Fréchet differentiability at dense sets of points of continuous convex functions on separable reflexive spaces.* S. L. Troyanski made a crucial step forward when he proved that *every WCG space can be renormed by an LUR norm* (see, e.g., [FHHMZ11, p. 587]). S. L. Troyanski and, independently, J. Lindenstrauss, showed that ℓ_∞ *does not admit any LUR norm* (see, e.g., [DeGoZi93b], [FHHMZ11, p. 409], and [HMVZ08]). M. I. Kadets and S. L. Troyanski asked if conversely, every space that admits no LUR norm must contain a copy of ℓ_∞. This question was first negatively answered in [HayZi89]; this paper originated from a J. Lindenstrauss' suggestion. Now is known that there is a space that does not contain a copy of ℓ_∞ and does not admit any strictly convex renorming [Hay99].

In [HMZ12, Sect. 5] several similar notions regarding some type of uniformity in convexity are collected, and known relationships among them are also presented, together with a list of related problems. In this section we repeat one of the problems there and add to that list a few more.

Problem 100 (Problem 15, [HMZ12]). Is it true that a separable Banach space X does not contain a copy of ℓ_1 if, and only if, X can be renormed by a norm that satisfies the following: whenever a sequence $\{x_n\}_{n=1}^\infty$ in S_X satisfies $\lim_{n,m} \|x_n + x_m\| = 2$, then $\{x_n\}_{n=1}^\infty$ is weakly Cauchy.

The property of the norm in Problem 100 is called in [HMZ12] **weakly Cauchy rotundness** (**WCR**, in short). It is not known whether the property WCR of a norm is equivalent to the property W*CR (replace the weak by the w*-topology in the definition) of its bidual norm (see [HMZ12, Problem 14]).

It is known that reflexivity implies WCR renormability (via a W2R renorming; the norm of a Banach space X is **weakly 2-rotund** (**W2R**, in short) if given a sequence $\{x_n\}_{n=1}^\infty$ in B_X such that $\|x_n + x_m\| \to 2$ as $n, m \to \infty$, there exists $x \in X$ such that $x_n \to x$ in the weak topology; *a Banach space is reflexive if, and only if, it is weakly 2-rotund renormable* [HaJo04]).

A Banach space X is called an **Asplund space** if Y^* is separable for every separable subspace Y of X (see also Sect. 4.1 for an equivalent definition, in fact the original one).

The norm of a Banach space X is **weakly uniformly rotund** (**WUR**, in short) whenever $x_n - y_n \to 0$ in the weak topology as soon as $\{x_n\}_{n=1}^\infty$ and $\{y_n\}_{n=1}^\infty$ are sequences in B_X such that $\|x_n + y_n\| \to 2$. If X is separable, WUR renormability is equivalent to X being an Asplund space. See, e.g., [FHHMZ11, p. 387].

For further details, see, e.g., [HMZ12].

If a Banach space X is WUR, then so are its separable subspaces, and thus X is Asplund. The space JL_0 of Johnson and Lindenstrauss (see, e.g., [Ziz03]) is an example of an Asplund space that has no equivalent WUR norm.

Problem 101. Assume that X is a Banach space and Y is a subspace of X. Assume that both Y and X/Y admit equivalent WUR norms. Does X admit an equivalent WUR norm?

If the WUR property is replaced by the uniform convexity, the answer to Problem 101 *is positive,* see [EnLiPi75]. See also, e.g., [CasGon97].

The recent paper [SmiTr10] gives an account of the state of the art in LUR renormability for C(K)-spaces up to this date. It is mentioned there the following result of R. Haydon [Hay08]: *a Banach space X is LUR renormable whenever X* admits a dual LUR norm.* Problem 1 in [SmiTr10] is related to the following question. If X is a Banach space, the dual norm $\| \cdot \|$ on X^* is called w^***-LUR** if $x_n^* - x^* \to 0$ in the w^*-topology of X^* whenever x_n^*, x^*, $n \in \mathbb{N}$, are on the unit sphere and $\|x_n^* + x^*\| \to 2$.

Problem 102. Assume that X admits a norm the dual of which is w^*-LUR. Does X admit an LUR norm?

We took this problem from [Hay08, p. 2026].

It is known that *a Banach space X admits a norm whose dual norm is w^*-LUR if and only if the dual unit ball endowed with the w^*-topology is a descriptive compactum* [SmiTr10]. Therefore, Problem 102 could be stated as follows: does (B_{X^*}, w^*) descriptive compactum implies that X is LUR renormable?

There exists a conjecture that constitutes a particular case of problem 102 saying that $C(K)$ admits an LUR norm if K is descriptive. In [Haj98] it is shown that $C(K)$ has an equivalent LUR norm if K is a Namioka–Phelps compactum (these form a subclass of the class of descriptive compacta). An alternative approach to prove this can be found in [MMOT10]. For definitions and more on this matter, see also [SmiTr10].

The following problem is posed in [MOTV09].

Problem 103. Assume that the norm of a Banach space is strictly convex and that the weak topology on the unit sphere of X is metrizable. Does X admit an equivalent LUR norm?

The norm $\| \cdot \|$ of a Banach space is said to be **midpoint locally uniformly rotund** (**MLUR**, in short), if for every $x \in S_X$ and every sequence $\{x_n\}_{n=1}^\infty$ in B_X such that

$\|x + x_n\| \to 1$ and $\|x - x_n\| \to 1$, then $\|x_n\| \to 0$. Every LUR norm is MLUR, and certainly MLUR implies strict convexity. However, there are MLUR norms that are not LUR (see, e.g., [Sm81]).

Problem 104 (R. Haydon). Assume that X is an Asplund space that admits an equivalent strictly convex norm. Does X admit an equivalent MLUR norm?

We refer to [Hay99].

A norm $\| \cdot \|$ on a Banach space X is called **average locally uniformly rotund** (**ALUR**, in short) if each point x of S_X is an extreme point of the ball and the identity map from (B_X, w) onto $(B_X, \| \cdot \|)$ is continuous at x.

Problem 105. Let X be a separable Banach space. Is it true that X does not contain a copy of ℓ_1 if and only if X has an equivalent norm the dual of which is ALUR?

We refer to [La04], where a characterization of Banach spaces having a renorming with dual ALUR norm was provided. A later paper by the same author [La11] characterizes the same property in the case that the space does not contain a copy of ℓ_1.

There is a need for a comprehensive list of practical examples of norms with several different rotundity properties, in our opinion a good MSc project in this area.

2.9 More on the Structure of Banach Spaces

If X is a closed subspace of $C[0, 1]$ and $t_0 \in [0, 1]$, the point t_0 is called an **oscillating point for X** if there is $a > 0$ such that for every nondegenerated interval I around t_0, there is $x \in X$ such that the oscillation of x on I is greater than or equal to a. The set of all such points is called the **oscillation spectrum of X** and is denoted $\Omega(X)$.

Problem 106. In the notation above, assume that $\Omega(X)$ is countable. Is then X isomorphic to a subspace of c_0?

We took this problem from [EnGuSe14]. In this paper it is proved that *the answer to this problem is positive if we assume that $\Omega(X)$ is finite*.

Problem 107 (R. Aron, V. Gurarii). Is the subset of ℓ_∞ formed by all the elements that have only finite number of zero coordinates, spaceable in ℓ_∞?

Spaceability was defined in Problem 47.

We took this problem from [EnGuSe14]. In this paper, the following known results are discussed:

1. *The set M of all continuous functions on $[0, 1]$ that are nowhere differentiable is spaceable in $C[0, 1]$. In fact, $M \cup \{0\}$ contains a linear isometric copy of any separable Banach space.*
2. *The set of all differentiable functions on $[0, 1]$ is not spaceable in $C[0, 1]$.*
3. *The set of all continuous functions on $[0, 1]$ that are differentiable on $(0, 1)$ is spaceable in $C[0, 1]$.*

For references, see [EnGuSe14].

In this paper it is proved among other things that *for every infinite-dimensional subspace X of $C[0, 1]$, the set of functions in X that have infinite number of zeros is spaceable in X*.

Chapter 3
Biorthogonal Systems

In this chapter we review several problems on **biorthogonal systems** in Banach spaces, i.e., families $\{x_\gamma, f_\gamma\}_{\gamma \in \Gamma}$ in $X \times X^*$, where X is a Banach space, such that $\langle x_\alpha, f_\beta \rangle = \delta_{\alpha,\beta}$ whenever α and β belong to Γ. Here, $\delta_{\alpha,\beta} = 1$ if $\alpha = \beta$ and 0 otherwise. Note that Schauder bases, together with their functional coefficients, are examples of biorthogonal systems. The theory of biorthogonal systems is crucial for understanding the structure of Banach spaces, in particular of nonseparable ones. Many problems in this area are widely open. In the nonseparable case the theory of biorthogonal systems often goes as deep as to the roots of Mathematics, i.e., they use special axioms of Set Theory. In this respect we refer, for the most basic information, to, e.g., [HMVZ08, pp. 148 and 152] or [To06].

Here we will be touching on the following Martin's axiom MA_{ω_1}, consistent with ZFC, which, when translated in ZFC to topological terms, says: *Let K be a **CCC** compact space (i.e., does not contain any uncountable collection of pairwise disjoint open sets.) Let $\{U_\alpha\}_{\alpha < \omega_1}$ be a collection of open and dense sets in K. Then $\bigcap_{\alpha < \omega_1} U_\alpha$ is dense in K.* By considering the complements of singletons in $[0, 1]$, we see that this axiom, in the setting of Baire category, contradicts the Continuum Hypothesis. We will meet also Martin's MM axiom, which is stronger than MA_{ω_1} and yet consistent with ZFC. Martin's axioms, when using the so-called **root lemma** (see, e.g., [DeGoZi93, p. 262] and [FHHMZ11, p. 650]), provide an ideal environment for constructing uncountable biorthogonal systems as discovered by S. Todorčević (see, e.g., [HMVZ08, p. 153]). This is so much needed in the nonseparable theory.

3.1 Bases, Finite-Dimensional Decompositions

The original question formulated by S. Banach in 1932, and later by A. Pełczyński in 1964, was more precise that the version given in Problem 32: It was whether every infinite-dimensional Banach space had a separable infinite-dimensional quotient

© Springer International Publishing Switzerland 2016
A.J. Guirao et al., *Open Problems in the Geometry and Analysis of Banach Spaces*,
DOI 10.1007/978-3-319-33572-8_3

with a Schauder basis. This was solved in the affirmative *for separable Banach spaces* by W. B. Johnson and H. P. Rosenthal in 1972 (see, e.g., [LinTza77, p. 10] or [FHHMZ11, p. 195]). So, the Banach–Pełczyński problem mentioned here would have a positive answer as soon as Problem 32 had a positive answer.

The following question asks then for a possible strengthening of the Johnson–Rosenthal result mentioned above. This problem already appeared in [LinTza77, Problem 1.b.10].

Problem 108. Is it true that for every separable Banach space X there is a subspace Y such that both Y and X/Y are infinite-dimensional and have Schauder bases?

We refer to [FHHMZ11, p. 220]. The question is better understood in the setting of the so-called **three-space problems** in Banach spaces—i.e., asking whether a Banach space has property (P) as soon as a given subspace Y and the quotient X/Y both have property (P). For an exhaustive account of this subject the reader is encouraged to look at [CasGon97]. Due to the fact that *having a Schauder basis is not a three-space property* (W. Lusky [Lus85], see also [CasGon97, Sect. 7.2]), the previous question is of importance. In the aforementioned paper, W. Lusky proved that *the answer to Problem* 108 *is positive if X contains an isomorphic copy of c_0.* Let us note in passing that I. Singer proved that *there exists a Banach space with a Schauder basis and a subspace of it also having a Schauder basis and such that no Schauder basis of the subspace can be extended to a Schauder basis of the whole space* (see, e.g., [BorVan10, p. 35]).

It is a classical result, due to W. B. Johnson and H. P. Rosenthal (see [LinTza77, Theorem 1.g.2]) that *the answer to Problem* 108 *is also positive if we change "having a Schauder basis" by "having a finite-dimensional decomposition"* (see also [CasGon97, Sect. 7.1]).

Recall that a Schauder basis is called **monotone** if all the basis projections have norm exactly 1.

We cannot find in the literature the answer to the following problem:

Problem 109. Is it true that every infinite-dimensional Banach space contains a closed infinite-dimensional subspace with a monotone Schauder basis?

Note that F. Bohnenblust proved by probabilistic methods, for example, the important result that *there is a three dimensional Banach space that does not admit any monotone Schauder basis* [Boh41].

We refer to [FHHMZ11, p. 220] and to [Boh41].

The result of W. B. Johnson and H. P. Rosenthal mentioned in the Remarks to Problem 108 has a complement (see [LinTza77, Theorem 1.g.2]): *If X^* is separable, then both FDD's in Y and in X/Y can be chosen to be shrinking.* An FDD $\{X_n\}$ of a Banach space X (a concept defined in Problem 63) is said to be **shrinking** if $\|P_n^* x^* - x^*\| \to 0$ as $n \to \infty$, for every $x^* \in X^*$, where P_n is the n-th projection on X associated with the FDD, i.e., $P_n(x) = \sum_{i=1}^{n} x_i$ if $x = \sum_{n=1}^{\infty} x_n$, where $x_n \in X_n$ for all $n \in \mathbb{N}$. Note that if X has a shrinking FDD then X^* is separable.

The following seems to be an open problem:

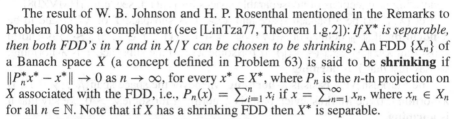

Problem 110. Assume that X is a separable Banach space such that X^* is nonseparable. Does there exist a subspace Y of X such that both Y and X/Y have nonseparable dual and both have finite-dimensional decompositions?

We refer again to [LinTza77, Sect. 1.g] and also to [DGHZ88], where *the result is proved for spaces that contain isomorphic copies of ℓ_1.* We note in passing that J. Hagler in [Ha87] showed that *if X is a separable Banach space with nonseparable dual, then X contains a subspace with Schauder basis and nonseparable dual.*

3.2 Markushevich Bases, Auerbach Bases

A biorthogonal system $\{x_\gamma, f_\gamma\}_{\gamma \in \Gamma}$ is called **fundamental in** X whenever $\overline{\text{span}}\{x_\gamma : \gamma \in \Gamma\} = X$, and is said to be **total** if $\bigcap_{\gamma \in \Gamma} \ker(f_\gamma) = \{0\}$. This second statement can be formulated by saying that the set $\{f_\gamma : \gamma \in \Gamma\}$ is **separating** in X^*, i.e., if $x \in X$, $x \neq 0$, then there is $\gamma \in \Gamma$ such that $f_\gamma(x) \neq 0$ (in other terms, the space span $\{f_\gamma : \gamma \in \Gamma\}$ is w^*-dense in X^*). The biorthogonal system $\{x_\gamma, f_\gamma\}_{\gamma \in \Gamma}$ is said to be **bounded** whenever a constant $C > 0$ exists such that $\|x_\gamma\|.\|f_\gamma\| \leq C$ for all $\gamma \in \Gamma$.

A **Markushevich basis** $\{x_\gamma, f_\gamma\}$ is a fundamental and total biorthogonal system in X. If $\{e_n\}_{n=1}^{\infty}$ is a Schauder basis in a Banach space X, and $\{e_n^*\}_{n=1}^{\infty}$ is the associated sequence of functional coefficients, the system $\{e_n, e_n^*\}_{n \in \mathbb{N}}$ is a Markushevich basis.

Note that the assumption that a space admits a Markushevich basis is much weaker than that of having a Schauder basis: In fact, *every separable Banach space has a Markushevich basis* (see, e.g., [LinTza77, p. 43] or [HMVZ08, p. 8]), while *there are separable Banach spaces without a Schauder basis* (see comments to Problem 52). The concept of a Markushevich basis is though strong enough to be used successfully in the study of the structure of separable and nonseparable spaces. For example, Markushevich bases are indispensable in the study of weak compact generating or injections of nonseparable spaces into $c_0(\Gamma)$, see, e.g., [HMVZ08].

Let $(X, \| \cdot \|)$ be a Banach space and $\lambda > 0$. A subspace Y of X^* is said to be λ-**norming** if $\|\| \cdot \|\|$ defined on X by $\|\|x\|\| := \sup\{|f(x)| : f \in Y, \|f\| \leq 1\}$ is a norm satisfying $\lambda\|x\| \leq \|\|x\|\| (\leq \|x\|)$. If for some $\lambda > 0$ the subspace Y is λ-norming, we say just that Y **is norming**.

The Markushevich basis is λ-**norming** whenever the space $\overline{\operatorname{span}}\{f_\gamma : \gamma \in \Gamma\}$ is λ-norming. If the Markushevich basis is λ-norming for some $\lambda > 0$, we say that it is **norming**.

Problem 111 (K. John).   Assume that X is a WCG Banach space. Does X admit a norming Markushevich basis?

Every separable Banach space has a 1-*norming Markushevich basis*, a result due to A. I. Markushevich himself (see, e.g., [HMVZ08, Theorem 1.22]). However, *there exists a WCG Banach space that does not admit any* 1-*norming Markushevich basis*. This is a result of S. L. Troyanski (see, e.g., [HMVZ08, Theorem 5.22]). *The space ℓ_∞ does not have any Markushevich basis* (W. B. Johnson, see, e.g., [FHHMZ11, p. 217]), although ℓ_∞ *is complemented in a Banach space with a Markushevich basis* (A. Plichko, see, e.g., [FHHMZ11, p. 217]). The existence of a Markushevich basis with some extra properties is often strong enough to ensure a good behavior of the space.

We refer to [HMVZ08, Sect. 5] for related results.

Problem 112 (G. Godefroy).   Assume that X is an Asplund space with a norming Markushevich basis. Is X necessarily WCG?

We refer to [HMVZ08, p. 211]. An attempt to solve this problem in the positive may require to study the so-called Jayne–Rogers selectors (cf., e.g., [DeGoZi93, p. 18]).

Recall that a Markushevich basis $\{x_\gamma, f_\gamma\}_{\gamma \in \Gamma}$ is called an **Auerbach basis** if $\|x_\gamma\| = \|f_\gamma\| = 1$ for all $\gamma \in \Gamma$. *Every finite-dimensional Banach space admits an Auerbach basis* (see, e.g., [FHHMZ11, p. 181], and [HMVZ08, Sect. 1.2]).

Problem 113 (A. Pełczyński). Does every separable Banach space admit an Auerbach basis? In particular, does the space $C[0, 1]$ admit an Auerbach basis?

Every infinite-dimensional Banach space contains an infinite-dimensional subspace with an Auerbach basis. This is a result of M. M. Day, see, e.g., [HMVZ08, p. 7].

A. Pełczyński and A. Plichko, independently, showed that *for every $\varepsilon > 0$ every separable Banach space admits a Markushevich basis $\{x_n, f_n\}_{n=1}^{\infty}$ such that* $\sup\{\|x_n\| . \|f_n\| : n \in \mathbb{N}\} < 1 + \varepsilon$. The paper [HajMon10] contains the proof of Plichko's statement that *every Banach space—not necessarily separable—that contains a Markushevich basis contains a bounded Markushevich basis.*

We refer to [HMVZ08, p. 14].

3.3 Weakly Compactly Generated Banach Spaces

Recall that a Banach space X is **weakly compactly generated**—WCG, in short— if there is a weakly compact set K in X such that X is the closed linear hull of K. Obviously, separable Banach spaces and reflexive Banach spaces are WCG. Indeed, for separable spaces X it is enough to take $K := \{(1/n)x_n : n \in \mathbb{N}\} \cup \{0\}$, where $\{x_n : n \in \mathbb{N}\}$ is a dense subset of S_X, to get a norm-compact linearly dense subset K of X, and for reflexive spaces X it is enough to take $K := B_X$. *The space $c_0(\Gamma)$, endowed with the supremum norm, is WCG for every nonempty set Γ* (inject in it the space $\ell_2(\Gamma)$). On the other hand, the space $\ell_1(\Gamma)$, for Γ uncountable, and ℓ_∞, are not WCG (see, e.g., [FHHMZ11, p. 576]). We refer, e.g., to [HMVZ08, Chap. 6] for the relationship between WCG spaces and biorthogonal systems. We may say that WCG Banach spaces and their relatives form the most important and pleasant class of nonseparable Banach spaces. The concept of weak compact generation was created in the 1960s of the last century by J. Lindenstrauss and his collaborators. The state of the art of WCG spaces is (partially) described in [DeGoZi93, HMVZ08], and [FHHMZ11].

The following problems are on the weak compact generating of spaces:

Problem 114. Is a Banach space X necessarily WCG if X^{**} is?

Note that

(i) *A space X may not be WCG and it may exist a WCG space Y such that $Y \supset X$*
    (H. P. Rosenthal). More precisely, H. P. Rosenthal proved in [Ro74] that *there is
    a probability measure μ such that $L_1(\mu)$* (that is clearly WCG, as follows from
    the fact that $L_2(\mu)$ injects in it) *contains a subspace (even with unconditional
    basis) that is not WCG.*

    Later, S.K. Mercourakis and E. Stamati showed in [MerSta06] that *assuming
    Martin's axiom, every WCG space of density less than c is **hereditarily WCG**,*
    i.e., all its subspaces are WCG.

    Note in passing that *every Asplund WCG space is hereditarily WCG* (see,
    e.g., [DeGoZi93, p. 245], [HMVZ08, p. 211], or [Ziz03, p. 1789]).

(ii) *A space X may not be WCG and at the same time have a WCG dual space.* This
    is a result of W. B. Johnson and J. Lindenstrauss in [JoLin74], the now called
    JL₂ **space**, see, e.g., [Ziz03, p. 1764].

For recent information on Problem 114 we refer to [ArMe05].

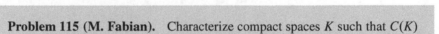

Problem 115 (M. Fabian). Characterize compact spaces K such that $C(K)$
is **hereditarily WCG**, i.e., each subspace of $C(K)$ is WCG.

We refer to [AviKa10].

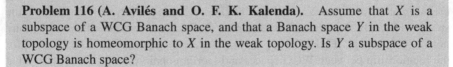

Problem 116 (A. Avilés and O. F. K. Kalenda). Assume that X is a
subspace of a WCG Banach space, and that a Banach space Y in the weak
topology is homeomorphic to X in the weak topology. Is Y a subspace of a
WCG Banach space?

We refer to [AviKa10].

At the birth of the WCG space theory it was conjectured [Lin72, Problem 9] that a
space with a Gâteaux differentiable norm must be a subspace of a WCG space. *The
conjecture was disproved* by W. B. Johnson and J. Lindenstrauss in [JoLin74].

Today it is known, e.g., that $C[0, \omega_1]$ *has a C^∞-Fréchet smooth norm and it is not a subspace of a WCG space* (see, e.g., [Ziz03, p. 1764]). *This space does not have any Y-lower semicontinuous equivalent Gâteaux smooth norm for Y being the norm closed linear hull of* $[0, \omega_1)$ *in* $C[0, \omega_1]^*$ [FMZ02, Proposition 8]. Thus, we pose the following problem:

Problem 117. Assume that a Banach space X admits an equivalent norm whose Y-lower semicontinuous envelope is Gâteaux differentiable for all norming subspaces Y of X^*. Is X necessarily a subspace of a WCG Banach space?

If Y is a norming subspace of X^*, then the norm $|x|_Y = \sup\{f(x) : f \in Y, \|f\| \leq 1\}$ is called the **Y-lower semicontinuous envelope** of the norm of X.

We note that *the converse implication in Problem 117 holds true* (see [FMZ02]). We refer to [FHHMZ11, p. 624] and to the aforementioned [FMZ02].

Chapter 4
Differentiability and Structure, Renormings

In this chapter we review some problems on smoothness, rotundity, and its connection to the structure of spaces. We recommend, for example, [BenLin00, DeGoZi93, Fa97, FHHMZ11, HMVZ08], and the recent book [HaJo14] for this area.

4.1 Asplund Spaces

The original definition of Asplund space was given in the following terms: a Banach space X is said to be **Asplund** if for every convex continuous function $f : X \to \mathbb{R}$ there exists a dense G_δ set $D \subset X$ such that f is Fréchet differentiable at every point $x \in D$. It was a pioneering result of J. Lindenstrauss that *every separable reflexive space is an Asplund space* [Lin63]. Later, E. Asplund gave another proof of this in [As68]. Still, the contribution of several mathematicians produced the following equivalence: *A Banach space X is Asplund if, and only if, every separable subspace Y of X has a separable dual* (see, e.g., [FHHMZ11, p. 486]).

There is a masterpiece-direct Baire category short proof of D. Preiss and L. Zajíček that *if X^* is separable, then X is an Asplund space* [PrZa84], see, e.g., [FHHMZ11, p. 414], or [BenLin00, p. 92].

A useful sufficient condition for Asplundness of a Banach space X is that X *admits a Lipschitz Fréchet differentiable bump function* (see, e.g., [FHHMZ11, p. 356]).

Problem 118 (G. Godefroy). Assume that all weak-star convergent sequences in the dual sphere S_{X^*} are norm convergent. Is then X an Asplund space?

© Springer International Publishing Switzerland 2016
A.J. Guirao et al., *Open Problems in the Geometry and Analysis of Banach Spaces*,
DOI 10.1007/978-3-319-33572-8_4

Of course, Problem 118 is a nonseparable problem only, since for separable Banach spaces the w^*-topology on the dual ball is metrizable and thus the assumption is that the weak-star and norm topologies on the dual sphere coincide. It is well known (cf., e.g., [FHHMZ11, p. 422]) that X^* is then separable. A positive solution of Problem 118 would provide an alternative proof of the Josefson–Nissenzweig theorem (we are indebted to G. Godefroy for the information). The Josefson–Nissenzweig theorem states that *in the dual sphere S_{X^*} of every infinite-dimensional normed space there is a sequence $\{x_n^*\}_{n=1}^\infty$ that w^*-converges to 0* (cf., e.g., [FHHMZ11, p. 151]).

It was proved in [BorFa93] that *the statement of the Josefson–Nissenzweig theorem is equivalent to the statement that on every infinite-dimensional Banach space there is a continuous convex function that is somewhere Gâteaux but not Fréchet differentiable.*

Thus we pose the following question:

Problem 119 (J. M. Borwein and M. Fabian). Can one show that every infinite-dimensional Banach space admits a convex continuous function that is somewhere Gâteaux but not Fréchet differentiable without a priori using the Josefson–Nissenzweig theorem?

This will give a new proof of this rather difficult result.

Problem 120. Assume that the norm $\| \cdot \|$ of a separable Banach space X has the property that the restriction of $\| \cdot \|$ to every closed subspace of X is somewhere Fréchet differentiable. Is X^* necessarily separable?

We took this problem from [GoMZ95].

We do not know if the following problem from [GoKal89] is still open:

Problem 121 (G. Godefroy and N. J. Kalton). Assume that X is a nonseparable Asplund space. Does X have an equivalent norm in which there is no proper closed 1-norming subspace in X^*?

For separable spaces the answer is positive. We remark that [GoKal89] contains many interesting open problems.

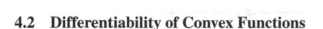

4.2 Differentiability of Convex Functions

The following problem relates to Alexandrov's finite-dimensional theorem (see [BorVan10, Theorem 2.6.4]).

Problem 122 (J. M. Borwein and D. Noll). Let f be a continuous convex function on a separable Hilbert space X. Do there exist $x, y \in X$ and a continuous symmetric bilinear form B on X such that for all $h \in X$ we have

$$f(x + th) - f(x) - (y, th) - t^2 B(h, h) = o(t^2)$$

as $t \to 0$?

We refer to [FHHMZ11, p. 511] and to [DeGoZi93, p. 176] and references therein for more information on this problem.

The following smooth variational principle is proved by using Baire category method: *let X be a Banach space that admits a Fréchet (Gâteaux) differentiable Lipschitz bump function. Then for every lower semicontinuous and bounded below proper real-valued function f on X and for every $\varepsilon > 0$ there exists a Fréchet (respectively, Gâteaux) differentiable Lipschitz function g on X such that* $\sup\{\|g(x)\| : x \in X\} \le \varepsilon$, $\sup\{\|g'(x)\| : x \in X\} \le \varepsilon$, and $(f - g)$ attains *its minimum on X* (see, e.g., [FHHMZ11, p. 355]). This is substantial in showing the Fréchet differentiability of convex functions on dense sets in Asplund spaces. However there is no version of this method for higher order differentiability on c_0, as c_0 *does not admit any uniformly Fréchet differentiable bump function*, see [FHHMZ11, p. 443]. Thus we have Problem 123.

Problem 123 (M. Fabian). Assume that f is a continuous convex function on c_0. Does there exist a point $x \in c_0$ such that

$$f(x+h) + f(x-h) - 2f(x) = O(\|h\|^2),$$

as $\|h\| \to 0$?

We took this problem from [Fa85] (see also [DeGoZi93, p. 177]). Note that for uniformly convex spaces the situation is different, see, e.g., [DeGoZi93, p. 162].

The following problem is so formulated to stress the evident need for further research in this area.

Problem 124. Let f_1, f_2, and f_3 be three Lipschitz real-valued functions on ℓ_2. Do they have a common point of Fréchet differentiability?

We remark that *the answer is "yes" for two such functions* and that *the answer is again "yes" if the functions are on the space c_0*. This is related to a celebrated theorem of D. Preiss [Pr90] stating that *every real-valued Lipschitz function on an Asplund space X is Fréchet differentiable at points of a dense set in X* (see, e.g., [LPT12]). We refer to [LPT12] and [Go01].

Let X be a Banach space. The space X is said to have the **convex point of continuity property** (**CPCP** in short) if every nonempty closed convex and bounded set in X has a point where the relative weak and norm topologies coincide. The space X^* is said to have the **convex* point of continuity property** (**C*PCP**, in short) if every w^*-compact convex set C in X^* has a point where the relative w^* and norm topologies coincide.

Problem 125. Let X be a separable Banach space such that X^* has the C*PCP. Is it true that every Lipschitz real-valued function on X that is Gâteaux differentiable everywhere is somewhere Fréchet differentiable?

We refer to [DGHZ87] and [DeGoZi93, p. 101] for more on this problem.

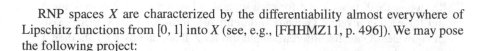

RNP spaces X are characterized by the differentiability almost everywhere of Lipschitz functions from $[0, 1]$ into X (see, e.g., [FHHMZ11, p. 496]). We may pose the following project:

> **Problem 126.** Can some characterization of CPCP spaces in this direction be found?

4.3   Smooth Extension of Norms

The norm $\| \cdot \|$ of a Banach space X is said to be **uniformly Fréchet differentiable** if the convergence of the difference quotient towards a continuous linear functional in the definition of a Fréchet differentiable norm is uniform on points of the sphere. It is equivalent to say that

$$\lim_{t \to 0+} \sup_{x, h \in S_X} \frac{1}{t}(\|x + th\| + \|x - th\| - 2) = 0.$$

A Banach space is superreflexive if, and only if, it admits such a norm (see, e.g., [FHHMZ11, Theorem 9.14]).

If $\lim_{t \to 0+} \sup_{x \in S_X} \frac{1}{t}(\|x + th\| + \|x - th\| - 2) = 0$ for each $h \in S_X$, we speak of **uniformly Gâteaux differentiable norm**.

> **Problem 127.** Assume that X is a reflexive (superreflexive) Banach space, Y is a subspace of X, and $\|\cdot\|$ is an equivalent Fréchet differentiable (respectively, uniformly Fréchet differentiable) norm on Y. Can $\|\cdot\|$ be extended to an equivalent Fréchet differentiable (respectively, uniformly Fréchet differentiable) norm on X? What about uniformly Gâteaux differentiable norms?

A similar problem for the case of Gâteaux differentiability has a negative answer. We refer to [DeGoZi93, p. 85] and [Ziz03, p. 1801].

Connected with Problem 127 let us formulate the following question:

Problem 128 (G. Godefroy).   Assume that a separable infinite-dimensional Banach space X is such that for each separable Banach space Z containing X and for each equivalent Gâteaux differentiable norm $\| \cdot \|$ on X there exists an equivalent Gâteaux differentiable norm on Z that extends $\| \cdot \|$. Is then X isomorphic to c_0?

This is connected to Zippin's theorem that reads: *A separable space X is isomorphic to c_0 if X is complemented in every separable overspace*. See, e.g., [Zip03].

A problem similar, at least formally, to Problem 128 is

Problem 129. Let X be a separable infinite-dimensional Banach space. Assume that for every pair Y, Z of isomorphic subspaces of X of infinite codimension, there is an automorphism τ of X so that $\tau Y = Z$. Is X isomorphic to c_0 or ℓ_2?

We refer to [LinTza77, p. 110]. An **automorphism** of a space is an isomorphism of the space onto itself.

4.4   Smooth Renormings

4.4.1   *Gâteaux Differentiability*

Recall that ω_1 is the first uncountable ordinal. The **long James space** $J(\omega_1)$ is the space of real-valued continuous functions f on the ordinal segment $[0, \omega_1]$ with $f(\omega_1) = 0$ for which the norm

$$\|f\| := \sup \left(\sum_{i=1}^{n} |f(\alpha_{i+1}) - f(\alpha_i)|^2 \right)^{\frac{1}{2}}$$

is finite, where the supremum is taken over all finite sequences $\alpha_1 < \alpha_2 < \ldots < \alpha_n$ of ordinals $\leq \omega_1$. We refer to [Edg79].

Problem 130 (R. Deville). Does the long James space $J(\omega_1)$ admit an equivalent Gâteaux differentiable norm or a continuous Gâteaux differentiable bump function?

Problem 131. Assume that K is a scattered compact such that $C(K)$ admits an equivalent strictly convex norm. Does $C(K)$ admit an equivalent Gâteaux smooth norm?

We refer to [MOTV09, p. 127].
Problems 104 *and* 131 *have a positive answer for trees* in [Smi07].

Problem 132. If K is the Kunen compact, does $C(K)$ admit a Gâteaux differentiable norm?

We refer to [SmiTr10, p. 409].

Problem 133. Assume that X is Asplund and admits an injection into $c_0(\Gamma)$, for some Γ. Does X admit a Gâteaux differentiable norm?

We refer to [OST, p. 220].

A compact space K is said to be a **Corson compact** if for some set Γ, K is homeomorphic to a compact subset C of $[-1, 1]^\Gamma$ in its pointwise topology, such that each point x of C has only a countable number of nonzero coordinates (we say that x is **countably supported**).

The class of Banach spaces having a dual unit ball that is a Corson compact when equipped with the w^*-topology has a number of interesting properties. Banach spaces belonging to this class are called **weakly Lindelöf determined** (WLD, in short). They are closely connected to the existence of PRIs in the space, see, e.g., [HMVZ08, Chap. 5].

An important subfamily of WLD spaces is formed by the so-called **weakly countably determined** or Vašák spaces: A Banach space X is called a **Vašák space** if there is a countable collection $\{K_n\}$ of w^*-compact subsets of X^{**} such that for every $x \in X$ and every $u \in X^{**} \setminus X$ there is n_0 for which $x \in K_{n_0}$ and $u \notin K_{n_0}$ [Vas81]. The sets K_n can be taken to be convex (see, e.g., [FHHMZ11, p. 609]). *Every subspace of a WCG space is Vašák.* Vašák spaces are especially useful in connection with smoothness of spaces.

For more on Vašák spaces we refer to [HMVZ08, DeGoZi93], and [Ziz03].

Problem 134. Let X be a Banach space. If X is a WLD space and if it admits a Lipschitz Gâteaux differentiable bump, does X admit a Gâteaux differentiable norm?

We refer to [FMZ06] and [Ziz03].

Problem 135 (S. A. Argyros and S.K. Mercourakis). Assume that X is a WLD Banach space. Is then every convex continuous function on X Gâteaux differentiable at points of a dense set in X?

We refer to [ArMe93], [Ziz03, p. 1800], and [HMVZ08, Chaps. 4 and 5] for more on this problem.

Problem 136.   Let K be a scattered compact.

(i) Does $C(K)$ admit a C^∞-smooth bump function?
(ii) If $C(K)$ admits a Fréchet differentiable norm, is it true then that $C(K)$ necessarily admits an equivalent C^∞-smooth norm?

This is true for metrizable K [FHHMZ11, pp. 626–627, 474]. We refer to [Ziz03, p. 1799] and [HaJo14, p. 326].

For an information on P. Hájek and R. Haydon partial positive answer for (ii) we refer to [HaJo14, p. 307].

If the dual norm of a Banach space $(X, \|\cdot\|)$ is strictly convex, then the norm $\|\cdot\|$ is Gâteaux differentiable. This is one of the results known as "Šmulyan test" (see, e.g., [FHHMZ11, p. 343]). The converse is not true: *the norm $\|\cdot\|$ can be Gâteaux differentiable and yet the dual norm fails to be strictly convex.* Examples of this situation were first given in [Klee59] and [Tro70]. It was proven in [GMZ12] that *if X is a nonreflexive subspace of a WCG Banach space, then there exists an equivalent LUR and Gâteaux differentiable norm on X such that its dual norm is not strictly convex. If X is, moreover, Asplund, then norm can even be chosen to be Fréchet differentiable, and such that the w and w^* topologies coincide on its dual unit sphere.*

We mentioned in notes after Problem 116 that $C[0, \omega_1]$ *admits a C^∞-Fréchet smooth norm.* However, M. Talagrand proved in [Ta86] that $C[0, \omega_1]$ *cannot be renormed by a norm whose dual would be strictly convex.* See [DeGoZi93, p. 313].

Problems 137 and 138 below address the question of the validity of the converse of the Šmulyan test under renorming, if the space is subjected to an extra hypothesis.

Problem 137 (S. L. Troyanski).   Assume that X has an uncountable uncon-ditional basis and admits a Gâteaux differentiable norm. Does X^* admit a dual strictly convex norm?

Problem 138.   Let X be a WLD Banach space. If X admits a Gâteaux differentiable norm, does X admit a norm the dual of which is strictly convex?

We refer to [FMZ06].

Problem 139. (i) Does there exist a C^2-smooth norm on c_0 such that its dual norm is not strictly convex? (ii) Does there exist on c_0 a C^2-smooth norm whose unit ball is dentable?

Let us note in passing that c_0 *does not have any C^2-smooth LUR norm* (see, e.g., [DeGoZi93, p. 187]), and that P. Hájek proved that *if Γ is uncountable, then $c_0(\Gamma)$ admits no C^2-smooth strictly convex norm* (see, e.g., [Ziz03, p. 1797]).

Problem 140. Based on Talagrand's result mentioned in the notes to Problem 137, find more scattered compacts K such that $C(K)^*$ does not admit a dual strictly convex norm.

Problem 141. Let X be a WCG space. Is it true that X admits an equivalent Gâteaux differentiable norm that is nowhere Fréchet differentiable if and only if X^* does not have the C^*PCP?

For separable spaces the answer is positive, see, e.g., [DeGoZi93, p. 101].

4.4.2 Strongly Gâteaux Differentiability

A norm is **strongly Gâteaux differentiable at** $x \in S_X$ if it is Gâteaux differentiable at x and whenever $f_n \in S_{X^*}$ are such that $f_n(x) \to 1$, $n \in \mathbb{N}$, then $\{f_n\}$ weakly converges to the derivative of the norm at x.

Problem 142. Find a characterization of separable Banach spaces that can be renormed by a strongly Gâteaux differentiable norm.

See [Tan96] and [HMZ12].

Problem 143. Find a characterization of the separable Banach spaces that can be renormed by a Gâteaux differentiable norm that is nowhere strongly Gâteaux differentiable.

See [Tan96] and [HMZ12].

4.4.3   Fréchet Differentiability

A function $f : U \to \mathbb{R}$, where U is an open subset of a Banach space X, is said to be C^1-**smooth** if it is Fréchet differentiable in U and the mapping $x \to f'(x)$ that to an $x \in U$ associates its Fréchet derivative (a continuous linear form on X) is continuous.

One of the most important problems in the area of Fréchet differentiability is the following one:

Problem 144. Does every Asplund Banach space admit a C^1-smooth bump function? In particular, does $C(K)$ admit a C^1-smooth bump function if K is the Kunen compact?

We refer to [Hay99]. See also, e.g., [FHHMZ11, p. 357], [HMVZ08, p. 151], and [HaJo14, p. 326] for more information on this problem.

Problem 145. Does a Banach space X admit a C^1-smooth bump function (norm) if there is a subspace Y of X such that both Y and X/Y admit C^1-smooth bump functions (respectively, norms)?

For a partial solution we refer to [DeGoZi90] and [JM97].

We do not know an answer to the following problem:

Problem 146. Assume that a Banach space X admits a Fréchet differentiable bump. Does X admit a Lipschitz, C^1-Fréchet smooth bump?

We refer to [DeGoZi93, p. 89] for more on this problem.

Problem 147. Assume that every quotient space of a given Banach space admits a Fréchet differentiable norm. Does X admit a norm whose dual on X^* is LUR?

We note in passing that *there are examples of Banach spaces X admitting a Fréchet smooth norm that has a quotient without this property* (see [Hay99]).

Problem 148. Does ℓ_1 admit an equivalent norm that is Fréchet differentiable everywhere outside a countable family of hyperplanes?

We refer to [BenLin00, p. 96], and note that J. Vanderwerff proved in [Van92] that *every $\aleph_0$-dimensional normed space admits an equivalent norm that is outside the origin everywhere Fréchet differentiable.*

We can ask the following:

Problem 149. Let Γ be an uncountable set and F be the normed space of all finite supported vectors in $\ell_1(\Gamma)$ equipped with the norm from $\ell_1(\Gamma)$. Does F admit a Fréchet smooth norm?

Problem 150. Assume that X is a separable Banach space that does not contain an isomorphic copy of ℓ_1. Is the bidual norm of X^{**} somewhere Fréchet differentiable?

For the James JT space this was solved in the positive by W. Schachermayer. We refer to [Ziz03, p. 1800].

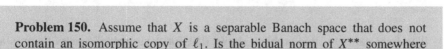

In [JoLin74], W. B. Johnson and J. Lindenstrauss posed the following problem that, despite many results in this direction, needs further research.

Problem 151. Characterize compact spaces K such that $C(K)$ admits an equivalent Fréchet differentiable norm.

We can ask the same problem for Gâteaux differentiable norms. In this direction we refer also to [SmiTr10].

Problem 152. Let X be a separable Banach space that does not contain an isomorphic copy of ℓ_1.

(i)   Does there exist an equivalent Gâteaux differentiable norm on X^*?
(ii) Is every w^*-compact set in X^* dentable?
(iii) Can X^* be equivalently renormed by a locally uniformly rotund norm?

We refer to [HaJo14, HMZ12, MOTV09], and [FHHMZ11, p. 511].

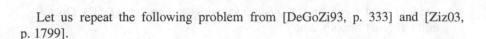

Let us repeat the following problem from [DeGoZi93, p. 333] and [Ziz03, p. 1799].

Problem 153. Let X be an Asplund space. Does X admit an equivalent Fréchet differentiable norm if and only if X admits an equivalent LUR norm?

The formulation of this problem shows that for "reasonable" spaces there is a self-duality of some properties, as we meet it in reflexive spaces or superreflexive spaces. We point out again that *X admits an equivalent LUR norm if X^* admits a dual LUR norm* [Hay08].

Problem 154. Assume that X is an Asplund space with the RNP. Does X necessarily admit a Fréchet smooth norm?

We took this problem from [DeGoZi93, p. 333].

Problem 155. Assume that a Banach space X admits a Fréchet smooth norm. Is the set of Fréchet smooth norms on X residual in the metric of uniform convergence on the ball of X?

We took this problem from [DeGoZi93, p. 90]. *The problem has a positive answer if X admits a norm whose dual is LUR* [DeGoZi93, p. 53] and *has a negative answer for higher order differentiability* (see [DeGoZi93, p. 189]). Note that *if X is an infinite-dimensional separable Banach space, then the set of all equivalent LUR norms on X in this metric is not Borel* [BGK96]. A similar situation is with Fréchet smooth norms if X^* is separable infinite-dimensional.

We do not know the answer to the following two problems:

Problem 156. Let X be a Banach space. Assume that the set of Fréchet smooth norms is dense in the space of equivalent norms on X in the metric of uniform convergence on the unit ball of X. Does X admit an equivalent norm the dual norm of which is LUR on X^*?

⎯⎯⎯⎯⎯◍⎯⎯⎯⎯⎯

Problem 157. Assume that a Banach space X admits a Fréchet smooth norm and also admits an LUR norm. Does X admit a norm that has both properties?

If X admits an LUR norm and X admits as well a norm whose dual is LUR, then X admits an LUR norm whose dual is LUR (see [DeGoZi93, p. 55]).

⎯⎯⎯⎯⎯◍⎯⎯⎯⎯⎯

Problem 158 (G. Godefroy).   Assume that X is an Asplund space. Does X admit a continuous (not necessarily equivalent) Fréchet differentiable norm?

Note that *the Banach space m_0 of countably supported bounded functions on $(0, \omega_1)$, where ω_1 is the first uncountable ordinal, endowed with the supremum norm, does not admit any continuous Gâteaux differentiable norm* (R. Haydon, see, e.g., [DeGoZi93, p. 89]).

⎯⎯⎯⎯⎯◍⎯⎯⎯⎯⎯

Given a convex function on a Banach space X, $\varepsilon > 0$, and a set $M \subset B_X$, we say that f is ε-M-**differentiable at** $x \in X$ if

$$\lim_{t \to 0} \frac{1}{t} \sup\{f(x + th) + f(x - th) - 2f(x) : h \in M\} < \varepsilon.$$

We say that the norm $\| \cdot \|$ of a Banach space X is σ-**Fréchet differentiable** if for every $\varepsilon > 0$ there is a decomposition

$$B_X = \bigcup_{n=1}^{\infty} A_n^{\varepsilon}$$

such that, for each $n \in \mathbb{N}$, the norm $\| \cdot \|$ is $\varepsilon\text{-}A_n^{\varepsilon}$-differentiable at every nonzero $x \in X$.

We say that the norm $\| \cdot \|$ of a Banach space X is **uniformly Gâteaux differentiable** if for every $h \in X$,

$$\lim_{t \to 0} \frac{1}{t} \sup\{f(x + th) + f(x - th) - 2f(x) : x \in S_X\} = 0.$$

It follows that *every uniformly Gâteaux differentiable norm is σ-Fréchet differentiable*. Thus the notion of σ-Fréchet differentiability provides a roof over both Fréchet and uniformly Gâteaux differentiability. We refer to [FMZ07].

Problem 159. Characterize $C(K)$-spaces that admit σ-Fréchet differentiable norm for K being an Alexandrov compactification of a tree space.

For the definition of tree spaces we refer, e.g., to [DeGoZi93, p. 266].

4.5   Strongly Subdifferentiable Norms

A norm $\| \cdot \|$ on a Banach space X is said to be **strongly subdifferentiable (SSD, in short) at a point** x if the one-sided limit

$$\lim_{t \to 0^+} \frac{1}{t}(\|x + th\| - \|x\|)$$

exists uniformly for $h \in S_X$.

A result of G. Godefroy (see, e.g., [GoMZ95] or [Go00, Proposition 8]) is that *every Banach space with an SSD norm is an Asplund space*. In the aforementioned paper [GoMZ95] we asked whether, in the other direction, every Asplund space admits an equivalent SSD norm. In [Go00, Theorem 9] it was proved that *if a Banach space X has an SSD norm, then every bounded and weakly closed subset of X is an intersection of finite unions of balls* and, as a consequence [Go00, Corollary 13], that *if X is a Banach space with an SSD norm, then X^* contains no proper norming subspace*. M. Jiménez and J. P. Moreno, in [JM97], constructed examples of *Asplund spaces without equivalent Fréchet smooth norms,* and G. Godefroy, by using the previous corollary, observed in [Go00] that, in fact, *those spaces do not have an equivalent SSD norm*. This answered in the negative our question in [GoMZ95]. Since those spaces are constructed under CH, the following remains, apparently, open:

Problem 160 (G. Godefroy). Does there exist an example in the sole ZFC of an Asplund space with no equivalent SSD norm?

For more information on the subject see [GoMZ95] and [Go00].

The separable non-Asplund Banach spaces are characterized as those separable Banach spaces admitting an equivalent norm that is nowhere SSD except at the origin [GoMZ95]. It is unknown whether the result holds true in the nonseparable case. So we formulate the following question [GoMZ95]:

Problem 161. Assume that X is a nonseparable non-Asplund Banach space. Does X admit an equivalent norm that is SSD only at the origin?

This problem has a positive answer if the space $(X, |\cdot|)$ has the SCP. Indeed, let Y_0 be a separable subspace of X with nonseparable dual and let $Y \supset Y_0$ be a separable complemented subspace in X, so there exists a projection P from X onto Y. Then Y^* is nonseparable. Thus, there is an equivalent norm $\|\cdot\|$ on Y that is nowhere SSD but at the origin. Thus, the norm $\|\|x\|\| := \|Px\| + |x - Px|$, where $P : X \to Y$ is a projection, is a nowhere (except at the origin) SSD equivalent norm on X.

The norm $\|\cdot\|$ of a Banach space is Fréchet differentiable at some $x \in X$ if, and only if, it is both SSD and Gâteaux differentiable at x. So it is natural to formulate the following question:

Problem 162. Assume that X admits an equivalent SSD norm and that X also admits an equivalent Gâteaux differentiable norm. Does X admit an equivalent Fréchet differentiable norm?

We took this problem from [GoMZ95].

4.6 Measure-Null Sets in Infinite Dimensions

As infinite-dimensional Banach spaces are not locally compact, difficulties arise with the concept of measure-null sets on that setting.

There are several definitions of measure-null sets in separable infinite-dimensional Banach spaces (a good reference is [BenLin00, Chap. 6], where an excellent and up-to-date treatment is presented).

We will just briefly discuss only two of their definitions. They deal with infinite-dimensional separable Banach spaces. In the finite-dimensional case, these concepts coincide with Lebesgue measure-zero sets.

1. Following J. P. R. Christensen [Chr72] we call, as now used in the literature, a Borel set A in an infinite-dimensional separable Banach space X a **Haar null set** if there is a Borel probability measure μ on X (associated with A) such that $\mu(A + x) = 0$ for every $x \in X$.

   To give an example, assume that there is a line L in X such that all its translates intersect the Borel set A in a set of one-dimensional Lebesgue measure zero. Then any probability measure on L equivalent to the Lebesgue measure on L can be used to get such a measure μ.

2. A Borel set A in an infinite-dimensional separable Banach space X is called an **Aronszajn null set** if for every sequence $\{x_n\}_{n=1}^{\infty} \subset X$ with a dense linear span in X, the set A can be represented as a countable union $A = \bigcup_{n=1}^{\infty} A_n$ such that, for every $n \in \mathbb{N}$, every line in the direction of x_n meets the set A_n in a set of Lebesgue measure zero.

   There is a very nice proof following this approach in [Arsz76, p. 155] of the fact that *every compact space in an infinite-dimensional separable Banach space is of Haar measure zero*. The intersections with lines are all at most singletons in this case.

   This way it can be shown that the non-compact unit sphere of the infinite-dimensional Hilbert space is of Haar measure zero.

It is proved in the aforementioned Chap. 6 of [BenLin00] that *every Aronszajn null set is a Haar null set but not vice versa.* In fact Hilbert's cube $C = \{x \in \ell_2 : 0 \leq x_n \leq \frac{1}{n}\}$ is a compact set that is not Aronszajn null (see [Arsz76] and [BenLin00, pp. 128 and 142]).

N. Aronszajn [Arsz76] and independently P. Mankiewicz [Man73] proved that *every Lipschitz real-valued function on a separable Banach space is Gâteaux differentiable outside an Aronszajn null set.* However, D. Preiss and J. Tišer showed in [PrTi95] that *every infinite-dimensional separable Banach space admits a real-valued Lipschitz function that is Fréchet differentiable only at the points of an Aronszajn null set.* E. Matoušková proved [Mat99] that *every separable infinite-dimensional superreflexive Banach space admits an equivalent norm that is Fréchet differentiable only at the points of an Aronszajn null set.*

We can pose the following problem:

Problem 163 (E. Matoušková). (i) Can Matoušková's result above be extended to some nonsuperreflexive spaces? (ii) Can Matoušková norm be made Gâteaux differentiable?

It was proved by E. Matoušková and C. Stegall [MatSt96] that *if X is an infinite-dimensional separable nonreflexive Banach space, then X contains a closed convex set with empty interior that is not Haar null.* E. Matoušková proved in [Mat97] that *if X is an infinite-dimensional separable superreflexive Banach space then every closed convex set with empty interior in X is a Haar null set.*

The following seems to be an open problem:

Problem 164 (E. Matoušková). If X is an infinite-dimensional separable Banach space that is reflexive but not superreflexive, can X contain a closed convex set with empty interior that is not Haar null?

We took this problem from [BenLin00, p. 162].

It is known that *Aronszajn null sets can be mapped by a Lipschitz equivalence onto sets that are not Aronszajn null.* It is also known that *Haar null sets can be mapped by a Lipschitz equivalence onto sets with Haar null complement.*

The following is apparently open:

Problem 165. Can an Aronszajn null set be mapped by a Lipschitz equivalence onto a set whose complement is Aronszajn null?

We took this problem from [BenLin00, p. 168].

4.7 Higher Order Differentiability

A function $\varphi : X \to \mathbb{R}$ is said to be **twice Gâteaux differentiable at** $x \in X$ provided that $\varphi'(y)$ exists in the Gâteaux sense for y in a neighborhood of x, that the limit

$$\varphi''(x)(h, k) = \lim_{t \downarrow 0} \frac{1}{t}(\varphi'(x + tk) - \varphi'(x))(h)$$

exists for each $h, k \in X$, and that $\varphi''(x)$ is a continuous symmetric bilinear form.

It was shown in [BorNo94, Proposition 2.2] that *if a norm is twice Gâteaux differentiable at x, then it is Fréchet differentiable at x*. On the other hand there are twice Gâteaux differentiable bump functions on $\mathbb{R}^2$ that are not Fréchet differentiable at some points. We refer to [Van93].

Problem 166 (J. Vanderwerff). Does there exist a Banach space that admits a continuous twice Gâteaux differentiable bump function but admits no Fréchet differentiable norm?

Problem 167. Assume that a Banach space X admits a twice Gâteaux differentiable norm and does not contain a copy of c_0. Is X necessarily superreflexive?

We refer to [HaJo14, p. 323].

Recall that a function $f : U \to \mathbb{R}$, where U is an open subset of a Banach space X, is said to be C^1-**smooth** if it is Fréchet differentiable in U and the mapping $x \to f'(x)$ that to an $x \in U$ associates its Fréchet derivative (a continuous linear form on X) is continuous. Sometimes it is possible to iterate this procedure—getting a mapping from U into the space of symmetric bounded n-linear forms endowed with the supremum norm. The eventual continuity of such a mapping defines the so-called C^n-**smoothness** of the original function f at a given point. The C^n-**smoothness** at U means C^n-smoothness at each of its points. Finally, C^∞-**smoothness** means the possibility to iterate the procedure with no end. Note that a useful *sufficient condition for the superreflexivity of a space X is that X admits a C^2-smooth bump function and does not contain a copy of c_0* (see, e.g., [DeGoZi93, p. 203]). For details see, e.g., [FHHMZ11, Sect. 10.1].

Problem 168. Assume that a Banach space X does not contain an isomorphic copy of c_0 and admits an equivalent norm the second derivative of which is locally uniformly continuous on the sphere of X. Does X admits an equivalent uniformly convex norm the second derivative of which is uniformly continuous on its sphere?

We refer to [DeGoZi93, p. 203].

Problem 169. Assume that a nonseparable WCG Banach space X admits a C^∞-smooth norm. Is every equivalent norm on X approximable by C^∞-smooth norms uniformly on the ball of X?

A positive answer to this problem is known for separable spaces, we refer to [HaJo14, p. 464]. It has recently been shown that *the answer is yes for the space* $c_0(\Gamma)$, *for any* Γ [BiSm16].

The following is a particular problem that may be useful in applications.

Problem 170. Let X be the space c_0 with an equivalent norm $\| \cdot \|$, $\{e_i\}_{i=1}^{\infty}$ be the sequence of standard unit vectors in c_0, and $\varepsilon > 0$. Does there exist an equivalent norm $\| \cdot \|_S$ on c_0 such that $\big| \|x\| - \|x\|_S \big| < \varepsilon$ for all x in the unit ball of c_0 and that $\inf_{i \in \mathbb{N}} \big| \|e_i\| - \|e_i\|_S \big| = 0$?

Problem 171. Is every continuous function on c_0 uniformly approximable by **real analytic** functions (i.e., functions admitting locally power series expansion)?

For a partial positive answer for uniformly continuous functions (a result due to R. Fry, M. Cepedello-Boiso, and P. Hájek) we refer to [HaJo14, p. 462]. We note that *for the space ℓ_2, Problem 171 has a positive solution*. This is J. Kurzweil's pioneering result (see, e.g., [HaJo14, p. 403]).

Problem 172. Is every equivalent norm on ℓ_2 approximable, uniformly on the ball of ℓ_2, by real analytic norms?

We took this problem from [HaJo14, p. 464].

Problem 173. Let X be a separable Banach space. Does X admit an equivalent C^{∞}-differentiable norm if X satisfies one of the following conditions?

(i) (A. Pełczyński) For each $k \in \mathbb{N}$, X admits an equivalent C^k-differentiable norm.

(ii) (H. P. Rosenthal) X admits a **separating polynomial**, i.e., a polynomial P such that $P(0) = 0$ and $\inf_{S_X} P > 0$.

For definition and properties of polynomials on Banach spaces we refer to [HaJo14], and for this problem to [HaJo14, pp. 322–323].

Problem 174. Assume that a separable Banach space X admits a C^k-differentiable bump function for $k > 1$. Does X admit an equivalent C^k-differentiable norm? Or, does X at least contain an infinite-dimensional subspace that admits an equivalent C^k-differentiable norm?

We refer to [HaJo14, p. 323].

Problem 175. Assume that a separable Banach space admits a bump whose second Fréchet derivative is uniformly continuous. Does X admit a norm whose second Fréchet derivative is uniformly continuous on the sphere?

We refer to [HaJo14, p. 323].

Problem 176. Assume that a separable Banach space admits a norm whose second Fréchet derivative is uniformly continuous on the sphere. Does X contain an isomorphic copy of some ℓ_p, $1 < p < \infty$?

We refer to [HaJo14, p. 323].

Problem 177 (A. Pełczyński). Assume that a Banach space X admits an equivalent C^2-differentiable norm. Is it true that X has the **weak Banach–Saks property** (i.e., if a sequence $\{x_n\}_{n=1}^{\infty}$ in X satisfies $x_n \to 0$ weakly, then there is a subsequence $\{x_{n_k}\}_{k=1}^{\infty}$ of $\{x_n\}_{n=1}^{\infty}$ such that $\|\frac{1}{k}(x_{n_1} + x_{n_2} + \cdots + x_{n_k})\| \to 0$)?

Problem 178. Does the space of compact operators on ℓ_2 admit an equivalent real analytic norm?

A C^∞-smooth norm on this space was constructed by N. Tomczak-Jaegermann (cf., [HaJo14, p. 325]).

Let us mention in passing that P. Hájek showed that *for a compact space K, C(K) admits a real analytic norm if, and only if, K is countable*, see, e.g., [HaJo14, p. 307], and P. Hájek and S. L. Troyanski found *the first example of a separable Banach space that admits a C^∞-smooth norm but no real analytic norm*, see, e.g., [HaJo14, p. 321].

Therefore, *for a metrizable compact space K, C(K) admits a Fréchet smooth norm if, and only if, it admits a real analytic norm* (see, e.g., [FHHMZ11, pp. 626, 627]).

A real analytic norm on c_0 was first constructed in [FPWZ89]. A. Pełczyński showed in [Pe57] that $c_0(\Gamma)$ admits no real analytic norm if Γ is uncountable.

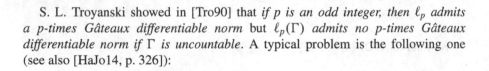

S. L. Troyanski showed in [Tro90] that *if p is an odd integer, then ℓ_p admits a p-times Gâteaux differentiable norm but $\ell_p(\Gamma)$ admits no p-times Gâteaux differentiable norm if Γ is uncountable*. A typical problem is the following one (see also [HaJo14, p. 326]):

Problem 179. Does ℓ_3 admit an equivalent four times Gâteaux differentiable norm or bump?

Some problems in nonseparable spaces heavily depend on the possibility to decompose the spaces into smaller spaces. This way the following problem has a positive solution in many cases (see, e.g., [DeGoZi93, Ziz03, HaJo14]) but in full generality is still open.

Problem 180. Assume that a Banach space X admits a C^k-smooth bump function (for $k \in \mathbb{N}$ or $k = \infty$). Is it true that any continuous mapping from X into a Banach space Y can be uniformly approximated by a C^k-smooth mapping?

Problem 181. Is it true that the set of all C^1-smooth norms is dense in the set of all the equivalent norms on the space $C[0, \omega_1]$ of continuous functions on the ordinal segment $[0, \omega_1]$, endowed with the topology of the uniform convergence on the unit ball?

We refer to [Ziz03, p. 1799] for more on this problem.

R. Fry proved in [Fry04] the following result: *Let X be a Banach space such that X^* is separable, G be an open set in X, and $F : G \to \mathbb{R}$ be a bounded and uniformly continuous map. Then for each $\varepsilon > 0$ there exists a C^1-smooth map K on G with bounded derivative and such that for all $x \in G$,*

$$|K(x) - F(x)| < \varepsilon.$$

Problem 182. Suppose that a Banach space X admits a C^k-smooth and Lipschitz bump function. Are Lipschitz mappings from X into a Banach space Y approximable by C^k-smooth and Lipschitz mappings?

We took this problem from [HaJo14, p. 464]. It is open even for X separable and Y general, or for X WCG and $Y := \mathbb{R}$.

What can be proved in this direction is the following result in [HaJo14, p. 428]: *Let X be a separable Banach space that admits a C^k-smooth Lipschitz bump function. Then there is a Lipschitz homeomorphism $\phi : X \to c_0$ such that the component functions $e_j^* \circ \phi$ belong to $C^k(X)$ for each j.*

Note that the latter result compares with the famous I. Aharoni result on Lipschitz homeomorphisms of separable Banach spaces into c_0 (see, e.g., [FHHMZ11, p. 546]). Note also that yet, ϕ may in general be nowhere Gâteaux differentiable, as the Gâteaux derivative will induce an isomorphism of X into c_0, something in general impossible (see, e.g., [FHHMZ11, Chap. 12]).

Problem 183. Let X be a separable Banach space that admits a C^k-smooth and Lipschitz bump function. Are Lipschitz real-valued functions on X approximable by C^k-smooth Lipschitz functions?

We refer to [HaJo14, p. 464] for more on this problem.

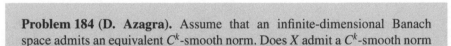

Problem 184 (D. Azagra). Assume that an infinite-dimensional Banach space admits an equivalent C^k-smooth norm. Does X admit a C^k-smooth norm in which it is not a complete space?

We refer to [Aza97]. Noncomplete norms are used in proving homeomorphisms with complements of some deleted sets in Banach spaces, a method that originated with C. Bessaga.

Problem 185 (D. Azagra and C. Mudarra). Let X be a Banach space that admits an equivalent C^k-smooth norm. Let $f : X \to \mathbb{R}$ be a Lipschitz and convex function and $\varepsilon > 0$ be given. Does there exist $\varphi : X \to \mathbb{R}$ that is C^k-smooth and convex and such that $|f - \varphi| \le \varepsilon$ on X?

We refer to [AzaMu15].

The following open question is taken from [AJR13].

Problem 186 (R. M. Aron, J. A. Jaramillo, and T. Ransford). Let Γ be a set of cardinality ω_1, $1 \le p < \infty$, and let F be a separable infinite-dimensional Banach space. Does there exist a C^∞-differentiable map $g : \ell_p(\Gamma) \to F$ such that

(continued)

Problem 186 (continued)
(i) g is surjective
(ii) when restricted to any separable subspace of $\ell_p(\Gamma)$, g is not surjective.

4.8 Mazur Intersection Property

In 1933, S. Mazur [Maz33] proved that *in any reflexive Banach space with a Fréchet differentiable norm, every convex bounded closed subset is an intersection of balls* (a Banach space that satisfies this property is said now to have the **Mazur Intersection Property** (**MIP**, in short)). Later on, his technique was combined with the Bishop–Phelps theorem to give that *the same result is true for any, non-necessarily reflexive, Banach space equipped with a Fréchet differentiable norm.*

Let us mention in passing a result of H. H. Corson and J. Lindenstrauss, who showed in [CorLin66] that *every weakly compact set in a Banach space is the intersection of a family of finite unions of balls*, see, e.g., [FHHMZ11, p. 418]. This result has then been further developed in [GoKal89].

Let us note in passing, too, that it is proved in [Ziz86] that *every WCG space can be renormed so that every weakly compact convex set is an intersection of balls*.

The MIP is a metric property, hence invariant under isometries, but not under isomorphisms. However, *if X admits such a norm, then the set of all such norms on X is residual in the metric of uniform convergence on the unit ball of X* [Geo91].

It was proved in [DeGoZi93h] that *if a Banach space X has the Radon–Nikodým property and admits a Fréchet differentiable bump, then it admits a norm with the MIP*. It is an open question whether the RNP requirement in the former statement can be removed. So we formulate the following:

Problem 187. Does every Banach space with a Fréchet differentiable bump have a renorming with the MIP?

Every Banach space with a Fréchet differentiable bump—in particular, with a Fréchet differentiable norm—is an Asplund space (see, e.g., [DeGoZi93, Theorem II.5.3]). R. R. Phelps proved in [Ph60] that *every* separable *Banach space has an MIP renorming if, and only if, it is an Asplund space*. R. Haydon [Hay90] provided an example of *an Asplund space without a Gâteaux renorming. This example can be renormed to have the MIP* [GJM04, Lemma 2.10], a positive answer to Problem VII.7 in [DeGoZi93]. It was natural then to ask—for a general Banach space—

whether there is a connection, if any, between being an Asplund space and having a renorming with the MIP. Any of the two suggested implications is false [JM97]: (1) *The non-Asplund space* $\ell_1 \oplus \ell_2(c)$ *can be renormed with the MIP.* (2) *Under CH, there are Asplund spaces that cannot be renormed to have the MIP* (an example is given by the space $C(K)$, where K is the Kunen compact). It has been proved that *the question of MIP renormability of Asplund spaces of density* ω_1 *is not decidable in ZFC* [BaHa08].

A result of R. Deville [De87] shows that *the long James space has an MIP renorming.* However, the existence of a Fréchet—or even Gâteaux—renorming of this space is an open question, as we saw in Problem 130.

———————◦◊◦———————

We can pose the following question, a particular case of Problem 187.

Problem 188. Assume that X admits a C^1-smooth bump function and does not contain a copy of c_0. Does X have an equivalent renorming with the MIP?

4.9 Krein–Milman and Radon–Nikodým Properties

A Banach space X is said to have the **Krein–Milman** property (**KMP**, in short) if every closed convex bounded subset of X has an extreme point. Note that this is the same as to require that every closed convex and bounded set in X is the closed convex hull of its extreme points (J. Lindenstrauss, see, e.g., [FHHMZ11, p. 378]).

The following is a famous known open old problem:

Problem 189 (J. Diestel). Assume that the Banach space X has the KMP. Is it true that X has the Radon–Nikodým property?

Note that the opposite implication is true, i.e., *every Banach space with the RNP has the KMP.* We refer to [FHHMZ11, p. 510]. Note, too, that, e.g., *for dual spaces Problem 189 was solved in the positive* by R. E. Huff and P. D. Morris [HuMo75].

There are some other cases in which the two properties are equivalent. A bounded convex subset D of a Banach space X is called **strongly regular** if for every convex subset C of D and every $\varepsilon > 0$ there are slices $S_1, \ldots, S_m$ of C such that

$$\mathrm{diam}\left(m^{-1}\sum_{i=1}^{m} S_i\right) < \varepsilon,$$

where diam denotes the diameter of a set. W. Schachermayer proved in [Sch87] that *strong regularity and KMP imply RNP of all bounded convex sets*. This gives, as a consequence, the aforementioned result of R. E. Huff and P. D. Morris. Another consequence is *the equivalence of KMP and RNP for spaces with an unconditional basis*.

Problem 190 (W. Schachermayer, A. Sersouri, and D. Werner). Assume that X fails the RNP and $\varepsilon > 0$ is given. Does there exist an equivalent norm on X such that each slice of the new unit ball has diameter greater than $2 - \varepsilon$?

We refer to [SSW89].

In the duality, we may formulate the following:

Problem 191 (G. Godefroy). Assume that X is a separable Banach space with nonseparable dual. Does there exist an equivalent norm $\|\cdot\|$ on X such that for every $x \in S_X$ and every $\varepsilon > 0$, there exists $y \in S_X$ with $\|x+y\| > 1-\varepsilon$ and $\|x-y\| > 1 - \varepsilon$?

We refer to [DeGoZi93, p. 122].

Problem 192. Assume that X has the Radon–Nikodým property. Does X admit an equivalent strictly convex or even LUR norm?

A. Plichko and D. Yost proved that *there is a nonseparable Banach space with the RNP that does not have the SCP* (however, *their space admits an LUR norm*) (see [HMVZ08, p. 167]). We refer, too, to [DeGoZi93, p. 177] and [MOTV09, p. 120].

We do not know if the following problem is still open:

Problem 193 (J. Bourgain). Assume that a separable Banach space X fails the RNP. Does there exist a subspace of X with Schauder basis that fails the RNP?

We took this problem from [Bou80].

A point x of a set S in a Banach space X is called an **exposed point of** S if there is $f \in X^*$ such that $f(x) = \sup_S f$ and $\{s \in S : f(s) = f(x)\} = \{x\}$.

Problem 194. Let X be a Banach space with the property that every closed convex bounded subset of X is the closed convex hull of its exposed points. Does X necessarily have the RNP?

We refer to, e.g., [HMZ12] and [Piet09] for more on this problem.

Problem 195 (P. Morris). Characterize the Banach spaces that can be renormed by an equivalent strictly convex norm $\| \cdot \|$ such that no point of the unit sphere S_X of the norm $\| \cdot \|$ is an extreme point of the unit ball of the double dual norm of X^{**}?

We refer to [Mo, GMZ15], and [GMZ14].

If a Banach space X does not contain a copy of c_0, then any closed convex bounded subset C of X contains a compact set K that is, in some sense, extreme in C, i.e., there is no $h \in X$, $h \neq 0$, such that $K \pm h \subset C$.

Problem 196. In the notation above, can the extremality of K be further developed for strong extremality, or exposedness, or so?

We refer to [FHHMZ11, p. 361].

Let X be a Banach space. A norm $\| \cdot \|$ on X is said to be **octahedral** if for every finite-dimensional subspace F of X and every $\eta > 0$, there exists $y \in S_X$ such that for every $x \in F$, we have

$$\|x + y\| \geq (1 - \eta)(\|x\| + 1).$$

G. Godefroy and B. Maurey proved that *a Banach space X contains an isomorphic copy of ℓ_1 if and only if X admits an equivalent octahedral norm* (see, e.g., [DeGoZi93, p. 106]).

Problem 197 (Wee-Kee Tang). Assume that X is a separable Banach space that contains an isomorphic copy of ℓ_1. Does X admit an equivalent Gâteaux smooth octahedral norm?

We refer to [HMZ12].

We recall that the **modulus of convexity** of a norm $\| \cdot \|$ is defined for $\varepsilon \in [0, 2]$ by

$$\delta(\varepsilon) = \inf\{1 - \|(x + y)/2\| : x, y \in S_X, \|x - y\| \geq \varepsilon\},$$

and the **modulus of smoothness** of a norm $\| \cdot \|$ is defined for $\tau > 0$ by

$$\rho(\tau) = \sup\left\{ \frac{\|x + \tau y\| + \|x - \tau y\|}{2} - 1 : x, y \in S_X \right\}.$$

If there exists $q > 0$, $K > 0$ such that

$$\delta(\varepsilon) \geq K\varepsilon^q$$

for every $\varepsilon \in (0, 2)$, then we say that the modulus of convexity of the norm is of **power type** q.

If there is $p > 1$, $K < \infty$ such that

$$\rho(t) \leq Kt^p$$

for all $t > 0$, we say that the modulus of smoothness of the norm is **of power type** p.

Problem 198. If a Banach space X admits a uniformly convex norm with modulus of convexity of power type p and admits a uniformly smooth norm with modulus of smoothness of power type q, does X admit a norm that simultaneously shares both these properties?

We refer to [DeGoZi93, p. 176] for more on this problem. *This problem has a positive answer for $p = q = 2$, see* [DeGoZi93, p. 175].

Problem 199. Assume that Γ is uncountable, f is a real-valued uniformly continuous function on the unit sphere of $c_0(\Gamma)$, and $\varepsilon > 0$ is given. Does there exist a nonseparable subspace of $c_0(\Gamma)$ such that on its unit sphere the oscillation of f is less than ε?

We note that *a separable version with an infinite-dimensional subspace instead* was proved by W. T. Gowers (see [BenLin00, p. 312]). This problem should be compared with Problem 17.

A norm $\| \cdot \|$ on X is called 2-**rotund** if whenever $\{x_n\}_{n=1}^{\infty}$ is a sequence in S_X such that $\lim_{n,m} \|x_n + x_m\| = 2$, then $\{x_n\}_{n=1}^{\infty}$ is convergent.

Problem 200 (V. D. Milman [Mil72]). Let X be a nonseparable reflexive space. Does it admit an equivalent 2-rotund norm?

We refer to [DeGoZi93, p. 177]. *For separable case, the answer is yes* (E. Odell and Th. Schlumprecht, see, e.g., [Go01]). A positive answer would provide for a full characterization of reflexive Banach spaces.

Milman's problem 200 had originally a second part asking whether every reflexive Banach space can be renormed by a **weakly 2-rotund** norm, i.e., a norm such that $\{x_n\}_{n=1}^{\infty}$ is weakly convergent whenever $x_n \in S_X$ are such that $\lim_{n,m} \|x_n + x_m\| = 2$. *This was solved in the positive* by P. Hájek and M. Johanis in [HaJo04].

Problem 201. Assume that X is a separable Banach space with a nonseparable dual. Does there exist a Gâteaux differentiable bounded Lipschitz function F from X into $\mathbb{R}^2$ such that

$$\|F'(x) - F'(y)\| \geq 1 \text{ for all } x, y \text{ in } S_X, x \neq y,$$

where the norm is meant in the space of all bounded linear operators from X into $\mathbb{R}^2$? The problem is open also for the James' tree space.

For the definition of the James' tree space see, e.g., [FHHMZ11, p. 233].

R. Deville and P. Hájek *proved in the positive Problem 201 in* [DeHa05] *for the space $X = \ell_1$.* For more on this problem see [DeIvLa15]. Note in passing that *if X and Y are separable and X is infinite-dimensional, then there is a continuous Gâteaux differentiable mapping f from X into Y with bounded support such that $f'(X) = B(X, Y)$* (i.e., the space of all bounded linear operators from X into Y) [ADJ03].

4.10 Norm-Attaining Functionals and Operators

In this section by a Banach space we mostly mean a real Banach space. Given a Banach space $(X, \|\cdot\|)$, the set $NA(X)$ consists of all the continuous linear functionals f on X that attain their norm, i.e., such that for some $x \in S_X, f(x) = \|f\|$.

R. C. James' weak compactness theorem asserts that *a Banach space X is reflexive if each continuous linear functional on X attains its maximum on B_X* (cf., e.g., [FHHMZ11, p. 137]).

The Bishop–Phelps theorem asserts that *for every Banach space X, the set of continuous linear functionals on X that attain their maximum on B_X is norm-dense in X^** (this solved a question posed by V. L. Klee; see [BiPhe61], see also, e.g., [FHHMZ11, p. 353]). We note that *the Bishop–Phelps theorem does not hold for incomplete normed spaces* ([Ph57], see, e.g., [Megg98, p. 271]).

In this connection, note that *if the norm of a Banach space X is LUR, then* it follows from the Šmulyan test that $NA(X)$ is a dense G_δ-subset of X^*.

If X is a separable space with a strictly convex norm, then $NA(X)$ is a Borel set in X^*. However, *every nonreflexive space can be renormed so that $NA(X)$ for the new norm is not a Borel set* [Kau91]. In this direction, we refer also to [Kur11].

The following two problems are in [DeGoZi93, p. 35], and we do not know if they are still open.

Problem 202. Assume that X is a separable Banach space. Denote by D the subset of S_{X^*} consisting of all support functionals to B_X at all points of Gâteaux differentiability of the norm of X. If D is norm separable, does it follow that X^* is separable?

A result of G. Godefroy [Go87] shows that this is the case if D is a $\|\cdot\|$-separable James boundary, since then the boundary is **strong** (i.e., $\overline{\mathrm{conv}}^{\|\cdot\|}(D) = B_{X^*}$). For the definition of a James boundary see comments to Problem 224.

Problem 203. If X is separable and X^* is nonseparable, does there exist a subset K of S_X, homeomorphic to the Cantor set, and $\varepsilon > 0$ such that if x and y are distinct points in K, and f and g are support functionals to B_X at x and y, respectively, then $\|f - g\| \geq \varepsilon$?

As mentioned in [DeGoZi93, p. 35], a positive answer to Problem 202 will imply a positive answer to Problem 203. Note that the closure of any uncountable subset of a separable complete metric space always contains a homeomorphic copy of the Cantor set.

Concerning the norm-attaining functionals, we would like to mention in passing the following sources and results:

1. P. Bandyopadhyay and G. Godefroy, in [BaGo06], proved among other things that *if X^{**} is separable, then X admits an equivalent norm in which $NA(X)$ is spaceable.*
2. C. A. De Bernardi and L. Veselý proved in [BerVe09] that *if K is a closed convex bounded set in a Banach space of dimension at least 2, then the set of all norm-1 supporting functionals to K is uncountable.*
3. The paper [AcoMon07] contains the proof that *every nonreflexive separable Banach space can be equivalently renormed so that $NA(X)$ for the new norm has an empty interior in the norm topology.* This solved a problem by I. Namioka. *The separability requirement was later removed* in the paper [AcoKa11].
4. Solving a problem of G. Godefroy, M. Rmoutil [Rm] recently proved that *there is a Banach space such that $NA(X)$ does not contain any 2-dimensional subspace.* Compare with item 1 above.

From the paper [BaGo06] we can formulate the following:

Problem 204. Does there exist a nonreflexive space such that $NA(X^*)$ is a linear subspace of X^{**}?

Note that $NA(c_0)$ is a dense linear subspace of ℓ_1.

The following problem appears in [AAAG07].

Problem 205. Does the spaceability of the set $X^* \setminus NA(X)$ characterize non-reflexivity?

The following problem is taken from [Aco06].

Problem 206. Is the subset of the norm-attaining 3-linear forms dense in the space of the 3-linear forms on $C(K)$ equipped with the usual supremum norm for the 3-linear forms?

We note that for general Banach spaces the answer is negative [AcAgPa96].

Regarding the set $S(D)$ of support functionals of a given bounded subset D of a Banach space, it is worth to mention the following result of J. Bourgain and Ch. Stegall (see, e.g., [Bour83, Theorem 3.5.5]): *Let C be a closed and convex subset of a Banach space X, and suppose that S(K) is of second category for each closed bounded convex subset K of C. Then C has the Radon–Nikodým property.* A closed bounded subset K of a Banach space X is said to have the **Radon–Nikodým property** (**RNP**, in short), if every closed convex subset of K is dentable. A closed and convex subset C of a Banach space is said to have the **Radon–Nikodým property** (**RNP**, in short), if each closed convex and bounded subset of C has the RNP.

The following question appears in [Bour83, Problem 3.5.6]:

Problem 207 (R. D. Bourgin). Does there exist a closed bounded convex set K in a Banach space X which is not dentable such that $S(K)$ if of second category?

As is mentioned in [Bour83, p. 58], the answer is "no" when K is separable, as it follows from the proof of the Bourgain–Stegall theorem mentioned above [Bour83, Theorem 3.5.5.]. M. Talagrand observed that the answer is also "no" when K is the closed unit ball of a space $C(T)$, where T is an infinite compact Hausdorff space.

The following problem concerns the validity of the Bishop–Phelps theorem in the context of complex Banach spaces (considering the attainment of the *modulus* of the complex linear functional). It is taken from [DeGoZi93, p. 35].

Problem 208. Let X be a complex c_0-space, and C be a bounded closed convex subset of X. Does there exist $f \in X^*$ such that the supremum of the modulus of f on the set C is attained? Is the set of all such f norm-dense in X^*?

V. Lomonosov proved in [Lo00] *that there is a complex Banach space X such that X^* is $\mathcal{H}^\infty$* (the space of bounded and holomorphic functions on the open unit disk, endowed with its usual supremum norm), *and a bounded, closed, and convex set $C \subset X$ such that the set of functionals attaining its maximum modulus at C is a 1-dimensional linear space. In fact, the closed convex hull of the set $\{\delta_z : z \in D\} \subset X^{**}$ given by $\delta_z(h) = h(z)$ for all $h \in \mathcal{H}^\infty$, where D is the open complex unit disk, is a subset C of X satisfying the above property.*

An operator from a Banach space X into a Banach space Y is said **to attain its norm** if there is $x \in S_X$ such that $\|Tx\| = \|T\|$. Denote by $NA(X, Y)$ the subset of $B(X, Y)$ consisting of all norm-attaining operators from X into Y.

In [BiPhe61], E. Bishop and R. R. Phelps raised the question of characterizing the Banach spaces X and Y for which the set $NA(X, Y)$ is dense in $B(X, Y)$. J. Lindenstrauss, in [Lin63], started the discussion of this general question by restricting its scope to characterizing which Banach spaces have property A—a space X has **property A** whenever $NA(X, Y)$ is dense in $B(X, Y)$ for every Banach space Y—and which Banach spaces have property B—a space Y has **property B** whenever $NA(X, Y)$ is dense in $B(X, Y)$ for every Banach space X. In the aforementioned paper, J. Lindenstrauss proved that *every reflexive Banach space has property A* (see also, e.g., [FHHMZ11, p. 359]). In fact, in the same paper [Lin63] he proved even more: *The set of all elements $T \in B(X, Y)$ such that $T^{**} : X^{**} \to Y^{**}$ attains its norm is always dense in $B(X, Y)$* (later on, in [Ziz73], this result was improved by showing that *the set of all elements $T \in B(X, Y)$ such that $T^* : Y^* \to X^*$ attains its norm is dense in $B(X, Y)$*). In [Lin63], J. Lindenstrauss gave examples of spaces not having property A (as $L_1(\mu)$ for a non-purely atomic measure μ, or $C(K)$, where K is an infinite compact metric space), as well as spaces not having property B (as a strictly convex space Y such that there is a non-compact operator from c_0 into Y). He even provided an example of a Banach space X such that $NA(X, X)$ is not dense in $B(X, X)$ (see also, e.g., [FHHMZ11, p. 424]). Lindenstrauss' result proving that every reflexive space has property A was later extended to RNP spaces by J. Bourgain in [Bou77], proving that *every RNP space (in any equivalent norm) has property A* (for a more precise statement, in fact a characterization of the RNP, see the notes to Problem 216). A later result of R. E. Huff [Hu80] shows that if X fails the RNP, then there exist two Banach spaces X_1 and X_2, both isomorphic to X, such that $NA(X_1, X_2)$ is not dense in $B(X_1, X_2)$.

Combined with Bourgain's result above, it follows that a Banach space has RNP if, and only if, for every Banach spaces X_1 and X_2 isomorphic to X, $NA(X_1, X_2)$ is dense in $B(X_1, X_2)$.

Lindenstrauss' paper [Lin63] turned out to be one of the most important papers in analysis on Banach spaces.

By using Troyanski's LUR renorming theorem for reflexive spaces (see, e.g., [FHHMZ11, p. 587]), in [Lin63] it was first showed that *continuous convex functions on reflexive Banach spaces are differentiable on dense sets* and *the balls of reflexive spaces are the closed convex hulls of their strongly exposed points* (a point $x \in B_X$ is **strongly exposed** if there is $f \in S_{X^*}$ such that f exposes x and $\|x_n - x\| \to 0$ whenever $x_n \in B_X$ are such that $f(x_n) \to 1$). These results influenced a huge part of the Banach space analysis since.

There are examples of a compact operator between Banach spaces that cannot be approximated by norm-attaining operators. This is a result of M. Martín [Mar14]. The following related problem is apparently open:

Problem 209. Can every finite-rank operator be approximated by norm-attaining operators?

With the definition of property B at hand, the fundamental result of E. Bishop and R. R. Phelps reads: $\mathbb{R}$ *has property B*. So it was natural to ask if this positive result can be extended to cover all finite-dimensional Banach spaces. The following problem appears explicitly in [JoWo79, Question 5]:

Problem 210. Does every finite-dimensional Banach space have property B?

The answer is positive for $(\mathbb{R}^n, \|\cdot\|_\infty)$. It is worth mentioning that W. Gowers proved in [Gow90] that *no infinite-dimensional Hilbert space has property B* (more generally, he proved in the aforementioned paper that ℓ_p *does not have property B for any* $1 < p < \infty$), and that later this result was extended by M. D. Acosta [Aco99] by proving that *no infinite-dimensional strictly convex Banach space has property B*.

Even the following particular case of Problem 210 (qualified in [JoWo79] as "the most irritating problem about norm-attaining operators which is linked to property B") is open:

Problem 211. Does $(\mathbb{R}^2, \|\cdot\|_2)$ have property B?

Problem 212. Characterize the topological compact spaces K such that $C(K)$ has property B.

A consequence of a result of W. Schachermayer [Sch83] is that $C[0, 1]$ *fails property B.*

Regarding property A, and in view of the Bourgain's result quoted above, it was natural to ask whether the RNP can be characterized by property A: so the following question seems to be open (see again [JoWo79, Question 3]).

Problem 213. Does property A imply the RNP?

This question is equivalent to whether property A is an isomorphism invariant, as it has been proved by R. E. Huff. The results of J. Bourgain and R. E. Huff mentioned in the introduction to Problem 210 conclude that *the RNP for a Banach space is equivalent to the fact that every renorming of X has property A.*

A question related to Problem 213 above is the following:

Problem 214. Let X be a Banach space without the RNP. Does there exist a renorming of X such that $NA(X, X)$ is dense in $B(X)$?

If X is isomorphic to Z ⊕ Z for some Banach space Z, the above question has a positive answer. This is a consequence of the Bourgain–Huff result mentioned above.

Problem 215 (M. Martín). Does a subspace of c_0 having the metric approximation property have property A?

J. Bourgain proved in [Bou77] that *a Banach space X has the RNP if, and only if, for every Banach space Y, and for every operator T from X into Y, for every bounded closed absolutely convex set B in X and every ε > 0, there is an operator T_1 from X into Y such that $\|T - T_1\| < \varepsilon$ and T_1 attains the maximum of the norm over B.*

Problem 216. Can the property CPCP be characterized in a similar way?

The following seems surprisingly to be an open problem.

Problem 217 (M. I. Ostrovskii). Does there exist an infinite-dimensional separable Banach space X such that every bounded linear operator from X into itself attains its norm?

We refer to [MaPl05].

There is close connection between the set of norm-attaining functionals and RNP-like properties of spaces. For a recent information, see, e.g., [GMZ] and references therein.

4.11 Weak Asplund Spaces

A Banach space X is called a **weak Asplund space** if every continuous convex function on X is Gâteaux differentiable at points of a residual set in X.

We note that *there is a non-weak Asplund space in which every continuous convex function is Gâteaux differentiable at the points of a dense set* [MoSu06]. Also, P. Holický, M. Šmídek and L. Zajíček proved that on every nonseparable reflexive Banach space there is a convex continuous function whose set of points of Gâteaux differentiability is not G_δ (though it must be residual) (see, e.g., [FHHMZ11, p. 357]).

Problem 218. Is $X \oplus \mathbb{R}$ a weak Asplund space if X is?

We refer to [Fa97, p. 48].

Problem 219. Is every subspace of a weak Asplund space weak Asplund itself?

For more on this problem see [Fa97, p. 170].

Problem 220. Assume that X is a weak Asplund space. Does X^* admit an equivalent strictly convex not necessarily dual convex norm?

The problem is mentioned on [Fa97, p. 100]. This should be compared with [FaGo88] where it is proved that *if X is an Asplund space, then X^* admits an equivalent not necessarily dual LUR norm*.

A topological space X is called **fragmented** if there is a metric ρ on X such that for every $\varepsilon > 0$ and every nonempty subset M of X there is an open set $\Omega \subset X$ such that the set $M \cap \Omega$ is nonempty and has ρ-diameter less than ε. If the dual ball B_{X^*} of a Banach space X is fragmented in its w^*-topology, then X is a weak Asplund space (see, e.g., [Fa97, p. 90]). However, the converse implication is not true [KMS01]. M. Fosgerau showed in [Fos92] that *a Banach space with a Lipschitz Gâteaux smooth bump has a weak-star fragmentable dual.* Thus, summing up, a weak Asplund space may not admit any Lipschitz Gâteaux smooth bump. However the following may possibly still hold:

Problem 221. Assume that the dual ball B_{X^*} of a Banach space X is fragmented in its w*-topology. Does X admit a Lipschitz Gâteaux smooth bump?

4.12 Polyhedral Spaces

A Banach space X is said to be **polyhedral** if the unit ball of each finite-dimensional subspace of X is a **polytope**, i.e., it is the convex hull of finitely many points. We refer to [FLP01, Go01a], and to [HaJo14, p. 290] for the basics on polyhedral spaces. Typically, *the space c_0 is polyhedral* (see, e.g., [Go01a]), and *every polyhedral space contains an isomorphic copy of c_0* (see, e.g., a nice proof in [FLP01, p. 658]).

We say that a real-valued function f on a Banach space X **locally depends on finitely many coordinates** if for every $x \in X$ there is a neighborhood U of x, a finite number $f_1, f_2, \ldots f_n$ of vectors in X^*, and a continuous real-valued function g on $\mathbb{R}^n$ such that for all $z \in U, f(z) = g(f_1(z), f_2(z), \ldots, f_n(z))$.

Problem 222. Assume that a separable Banach space X admits a continuous bump function that locally depends on finitely many coordinates. Is X isomorphic to a polyhedral space?

It is known that *if a separable space X admits a continuous bump that locally depends on finitely many coordinates, then X^* is separable and X is saturated with isomorphic copies of c_0* (see, e.g., [FHHMZ11, p. 472]).

It is known that *separable polyhedral spaces are characterized by the existence of an equivalent norm that locally depends on finitely many coordinates,* see [HaJo14, p. 278]. We mention here that, *for K the Kunen compact, the space $C(K)$ does not admit a continuous bump locally dependent on finitely many coordinates* (see, e.g., [Ziz03, p. 1798]). For more information on these questions we refer to [FHHMZ11, p. 468] and [HaJo14, p. 324].

Problem 223. Is every separable polyhedral space isomorphic to a subspace of an isometric predual of ℓ_1?

We refer to [HaJo14, p. 324].

<hr>

Problem 224 (H. Rosenthal). Assume that K is an infinite countable compact space. Is every quotient of $C(K)$ c_0-saturated?

It is not even known if a quotient of $C(\omega^\omega + 1)$ may contain an isomorphic copy of ℓ_2. We refer to [Ro03, p. 1571].

It is known (cf., e.g., [FHHMZ11, pp. 470–474]) that *every isomorphically polyhedral space is c_0-saturated*, and *it is equivalent, in the case of separable Banach spaces, (1) to admit a polyhedral norm, (2) to have a countable James boundary, and (3) to admit a norm depending on a finite number of coordinates.* A set $B \subset B_{X^*}$ is called a **James boundary** for a Banach space $(X, \|\cdot\|)$ if for every $x \in X$, $\|x\| = f(x)$ for some $f \in B$. Obviously, if K is a countable compact space, the space $C(K)$ has a countable James boundary (namely, the set $\{\pm\delta_k : k \in K\}$). We note, too, that the Hagler space [Ha77] is a separable non-Asplund space saturated by c_0 and thus it does not admit any continuous bump locally dependent on finitely many coordinates [FHHMZ11, p. 472]. An example of a non-isomorphically polyhedral space admitting a continuous bump locally dependent on finitely many coordinates is so far unknown (see Problem 222). Note that *there are polyhedral spaces with unconditional basis that have a space isomorphic to ℓ_2 as a quotient* (see, e.g., [HaJo14, p. 324]).

<hr>

Problem 225. Let X be a nonseparable Banach space. Is X polyhedral if and only if X admits a norm (respectively, a bump) depending locally on finitely many coordinates?

We refer to [FPST08, p. 452].

<hr>

Problem 226. Assume that a nonseparable Banach space X admits a norm that locally depends on finitely many coordinates. Does X admit a norm that is C^∞-smooth?

We refer to [Ziz03, p. 1799]. *For separable spaces this question has been solved positively* (P. Hájek, see, e.g., [FHHMZ11, p. 470]).

Problem 227. Does there exist a compact space K such that $C(K)$ admits a polyhedral norm but $C(K)$ does not admit a C^1-norm (or vice versa)?

We refer to [SmiTr10, p. 409].

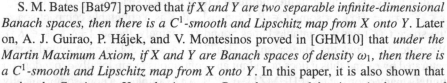

S. M. Bates [Bat97] proved that *if X and Y are two separable infinite-dimensional Banach spaces, then there is a C^1-smooth and Lipschitz map from X onto Y.* Later on, A. J. Guirao, P. Hájek, and V. Montesinos proved in [GHM10] that *under the Martin Maximum Axiom, if X and Y are Banach spaces of density ω_1, then there is a C^1-smooth and Lipschitz map from X onto Y.* In this paper, it is also shown that *under the Continuum Hypothesis, every Banach space of density c is the range of a C^∞-smooth and Lipschitz map from $c_0(c)$.* On the other hand, P. Hájek proved in [Haj98] that *there is no C^2-smooth map from c_0 onto any infinite-dimensional superreflexive space.* In [GHM10] the authors asked the following problems:

Problem 228 (A. J. Guirao, P. Hájek, and V. Montesinos).

1. Is it true that given a density α, all Banach spaces of this density are C^1-smooth images of each other?
2. Is it true that for $\alpha = c$, (1) holds even for C^∞-smooth mappings?

We refer to [FHHMZ11, p. 408] for more on this problem.

Chapter 5
Nonlinear Geometry

In this chapter we review several problems in the area of nonlinear structure of Banach spaces.

The topological classification of Banach spaces is fully understood: It was shown by M. I. Kadets [Ka67] that *all separable infinite-dimensional Banach spaces are mutually homeomorphic*—and, as shown by R. D. Anderson [And72], *they are homeomorphic to the product space* $\mathbb{R}^{\aleph_0}$; H. Toruńczyk proved that *all infinite-dimensional Banach spaces of the same density character are mutually homeomorphic* [Tor81]. Problems related to general nonlinear homeomorphisms will be treated in Sect. 5.4.

We will be dealing, besides linear (norm) isomorphisms, with nonlinear uniform homeomorphisms and nonlinear Lipschitz homeomorphisms of spaces. If no misunderstanding can occur we sometimes say "Lipschitz isomorphism" instead of "Lipschitz homeomorphism." A mapping $f : X \to Y$ between two metric spaces (X, d) and (Y, ρ) is said to be **Lipschitz** if there exists a constant C such that $\rho(f(x_1), f(x_2)) \leq Cd(x_1, x_2)$ for all $x_1, x_2 \in X$; a bijection φ from a Banach space X onto a Banach space Y is called a **Lipschitz (uniform) homeomorphism** if φ together with φ^{-1} are both Lipschitz (respectively, uniformly continuous).

While there are examples of nonseparable Banach spaces that are Lipschitz homeomorphic and not linearly isomorphic (see the notes before Problem 229 in page 105), we will see that it is an open problem if separable examples of this sort exist. There are examples of separable uniformly homeomorphic spaces that are not linearly isomorphic [Ri84] (N. J. Kalton proved in [Kal12] that *there are even two subspaces of c_0 that are uniformly homeomorphic but not linearly isomorphic*). We refer to [BenLin00, pp. 178, 244, 253, and 257].

Overall, we recommend the books [BenLin00] and [BePe75] in this area. Some results are discussed also in [FHHMZ11, Chap. 12]. We also refer to [Kal08] and the recent survey article [GoLaZi14].

© Springer International Publishing Switzerland 2016
A.J. Guirao et al., *Open Problems in the Geometry and Analysis of Banach Spaces*,
DOI 10.1007/978-3-319-33572-8_5

5.1 Lipschitz-Free Spaces

The following few problems deal with the concept of Lipschitz-free spaces (the definition is given below). They blend some linear and nonlinear aspects in Banach space theory.

Let M be a **pointed metric space**, that is, a metric space equipped with a distinguished point denoted by 0. The space $\mathrm{Lip}_0(M)$ is the space of all real-valued Lipschitz functions on M which vanish at 0. Equipped with the norm

$$\|f\| = \sup\left\{ \frac{|f(x) - f(y)|}{d(x,y)} : x,y \in M,\ x \neq y \right\},$$

where $f \in \mathrm{Lip}_0(M)$ and d is the metric of M, $\mathrm{Lip}_0(M)$ is a Banach space.

The Dirac map $\delta : M \to \mathrm{Lip}_0(M)^*$, defined by $\langle \delta(x), g \rangle = g(x)$, $x \in M$, $g \in \mathrm{Lip}_0(M)$, is an isometric embedding from M onto a subset of the dual Banach space $\mathrm{Lip}_0(M)^*$ considered in its dual canonical norm.

The norm-closed linear hull of $\delta(M)$ in $\mathrm{Lip}_0(M)^*$ is denoted by $\mathcal{F}(M)$, and is called the **Lipschitz-free space over** M. When M is separable, $\mathcal{F}(M)$ is separable.

We note that the unit ball of $\mathrm{Lip}_0(M)$ is compact in the topology of the pointwise convergence on M and thus also in the topology of the pointwise convergence on $\mathcal{F}(M)$. $\mathrm{Lip}_0(M)$ *is a dual space* and it follows that $\mathcal{F}(M)$ *is the natural canonical isometric predual of* $\mathrm{Lip}_0(M)$.

Banach spaces X are in particular pointed metric spaces (pick the origin as a distinguished point) and we can apply the previous construction. The isometric embedding δ defined above is of course nonlinear, since there exist Lipschitz functions on X which are not affine.

Although Lipschitz-free spaces over separable metric spaces constitute a class of separable Banach spaces which are easy to define, the structure of these spaces is not yet well understood.

If we identify (through the Dirac map) a metric space M with a subset of $\mathcal{F}(M)$, any Lipschitz map from M into a metric space N extends to a continuous linear map from $\mathcal{F}(M)$ to $\mathcal{F}(N)$. In this way Lipschitz maps become linear, but of course the complexity is shifted from the map to the free space. Namely, if M and N are two Lipschitz homeomorphic spaces, then $\mathcal{F}(M)$ and $\mathcal{F}(N)$ are linearly isomorphic. However, $\mathcal{F}(C[0,1])$ *is isomorphic to* $\mathcal{F}(c_0)$ (see [DuFe06] and [Kauf15]) and it is known that $C[0,1]$ *is not Lipschitz homeomorphic to* c_0 (see, e.g., [FHHMZ11, Theorem 12.61]).

It follows that $\mathcal{F}(\mathbb{R})$ is linearly isometric to L_1, but it is known that $\mathcal{F}(\mathbb{R}^2)$ *is not isomorphic to any subspace of* L_1 [NaSch07]. Thus, for a Banach space X, $\mathcal{F}(X)$ does not have the Radon–Nikodým property, and is not an Asplund space.

On the other hand, *if* $\mathbb{N}$ *denotes the set of the positive integers with the discrete metric, then* $\mathcal{F}(\mathbb{N})$ *is isomorphic to* ℓ_1 [Kal08].

Furthermore, let us mention that *if* X *is a separable Banach space, then* $\mathcal{F}(X)$ *contains a complemented copy of* X [GoKal03, Kal08]. More precisely, if X is a

Banach space, let β be a **barycenter map** from $\mathcal{F}(X)$ onto X, i.e., a map β with the property that $\beta\delta(x) = x$ for all $x \in X$. If X is separable, it is proved in [GoKal03] that *there is a linear isometry U from X into $\mathcal{F}(X)$ such that $\beta U = Id_X$.* Then $P := U\beta$ is a norm-1 projection from $\mathcal{F}(X)$ onto $U(X)$.

If X *is a nonseparable WCG space and Z_X is the kernel of β in $\mathcal{F}(X)$, then $Z_X \oplus X$ is Lipschitz isomorphic to $\mathcal{F}(X)$ but not linearly isomorphic to $\mathcal{F}(X)$* [GoKal03]. This gives a lot of examples of nonseparable spaces that are Lipschitz homeomorphic but not linearly isomorphic.

On the other hand, ℓ_∞ *is not isomorphic to a subspace of $\mathcal{F}(\ell_\infty)$* [GoKal03, Kal08].

If X *is a WCG space, then any weakly compact subset of $\mathcal{F}(X)$ is (norm) compact* [GoKal03].

The Lipschitz-free spaces are often called **Arens–Eells spaces**.

For more information on this subject we refer to [Kal08, GoKal03], and [Wea99].

Problem 229. Let M be an arbitrary **uniformly discrete metric space**, that is, there exists $\theta > 0$ such that $d(x, y) \geq \theta$ for all $x \neq y$ in M. Does $\mathcal{F}(M)$ have the bounded approximation property?

We refer to [GoOz14].

Note that *the approximation property holds in this case* [Kal04, Proposition 4.41]. A positive answer to Problem 229 would imply that every separable Banach space X is approximable (see Problem 64). On the other hand, a negative answer to this question would provide an equivalent norm on ℓ_1 failing the metric approximation property, and this would solve a famous problem in approximation theory, by providing the first example of a dual space—namely, ℓ_∞—equipped with the corresponding dual norm, having the AP (and even the BAP) but failing the MAP.

It is known that a Banach space X has the BAP if and only if $\mathcal{F}(X)$ *has the BAP* and that *there is a convex compact set K in a Banach space such that $\mathcal{F}(K)$ does not have the AP* [GoKal03] and [GoOz14]. Since every separable Banach space X is isometric to a complemented subspace of $\mathcal{F}(X)$, $\mathcal{F}(X)$ *fails the AP if a separable space X does.*

A. Dalet proved in [Dal15] that $\mathcal{F}(K)$ *has the MAP if K is a countable compact metric space.*

P. Hájek, G. Lancien, and E. Pernecká proved in [HLP] that *there is a compact space homeomorphic to the Cantor set C such that its free space does not have the AP.*

It is proved in [HaPe14] that $\mathcal{F}(\ell_1)$ *and* $\mathcal{F}(\mathbb{R}^n)$ *admit a Schauder basis.*

Problem 230. Find more separable Banach spaces X such that the space $\mathcal{F}(X)$ admits a Schauder basis.

We refer to [HaPe14]. See also Problem 231 below.

Problem 231. Let C be a closed subset of $\mathbb{R}^n$. Does $\mathcal{F}(C)$ admit a Schauder basis?

We refer to [HaPe14], where it is proved, for example, that *if C is a convex and bounded subset of $\mathbb{R}^n$, then $\mathcal{F}(C)$ has a Schauder basis.* In [Kauf15] it is showed that $\mathcal{F}(C)$ and $\mathcal{F}(\mathbb{R}^n)$ *coincide if C has moreover a nonempty interior*—hence $\mathcal{F}(C)$ has a Schauder basis in this case.

If we ask in Problems 230 and 231 for having the approximation property instead, the amount of available information is bigger (see [GoKal03] and [GoOz14]). For a related question, see in particular Problem 236 below.

Problem 232. Can $\mathcal{F}(\mathbb{R}^n)$ be isomorphic to $\mathcal{F}(\mathbb{R}^m)$ if $n \neq m$?

We refer to [Kauf15], where it is proved, for example, that *if $m < n$, then $\mathcal{F}(\mathbb{R}^m)$ is complemented in $\mathcal{F}(\mathbb{R}^n)$.*

Problem 233 (M. Cúth, M. Doucha, and P. Wojtaszczyk). Does $\mathcal{F}(\ell_2)$ or $\mathcal{F}(\ell_1)$ contain an isomorphic copy of c_0?

The problem is taken from [CDW]. This paper contains also the result that $\mathcal{F}(M)$ *is weakly sequentially complete whenever M is a subset of $\mathbb{R}^n$.*

Problem 234. Let C be some **cusp** in $\mathbb{R}^2$ (e.g., the union of the positive half x-axis with a positive parabola). Is $\mathcal{F}(C)$ linearly isomorphic to a subspace of L_1?

We refer to [Kauf15], where this question is proposed after proving some results on sets written as union of "orthogonal pieces."

Problem 235. Is it true that $\mathcal{F}(\ell_1)$ is complemented in its bidual?

If yes, Problem 243 below would have a solution in the positive. We refer to [GoLaZi14].

Problem 236. For which compact spaces K the space $\mathcal{F}(K)$ has the approximation property?

We refer to [GoOz14].

Problem 237. For which nonseparable Banach spaces X, the space $\mathcal{F}(X)$ admits a Markushevich basis? What about $\mathcal{F}(\ell_\infty)$?

Problem 238. For which nonseparable Banach spaces X, the space $\mathcal{F}(X)$ admits an LUR (respectively, strictly convex) norm? What about $\mathcal{F}(\ell_\infty)$?

Observe that $\mathcal{F}(X)$ is separable whenever X is separable. Therefore, the answer to both Problems 237 and 238 is affirmative in the separable setting.

5.2 Lipschitz Homeomorphisms

One of the most important open problems in this area is the following:

Problem 239. If X and Y are two Lipschitz homeomorphic separable Banach spaces, are they necessarily linearly isomorphic? The same question for nonseparable reflexive or superreflexive spaces.

We refer to [FHHMZ11, p. 560], [BenLin00, p. 169], and [Kal08, p. 22].

This should be compared with the classical Mazur–Ulam theorem that *any isometry from a Banach space X onto a Banach space Y that takes 0 to 0 is necessarily linear* [BenLin00, p. 341], or [FHHMZ11, p. 548].

I. Aharoni and J. Lindenstrauss proved in [AhaLi78] that *there is a nonseparable $C(K)$-space such that for some uncountable Γ the space $c_0(\Gamma)$ is Lipschitz homeomorphic to $C(K)$ but not linearly isomorphic to $C(K)$.*

It was proved by F. Albiac and N. J. Kalton in [AlKal09] that *there are two separable quasi-Banach spaces that are Lipschitz homeomorphic but not linearly isomorphic.*

A less challenging question regarding Problem 239 is perhaps the following:

Problem 240. Let X and Y be two Lipschitz homeomorphic separable infinite-dimensional Banach spaces. Does it exist an infinite-dimensional Banach space E that is linearly isomorphic to a subspace of X and at the same time isomorphic to some subspace of Y?

If X and Y are two Lipschitz homeomorphic Banach spaces and X is reflexive (respectively, superreflexive, having RNP, isomorphic to a Hilbert space), then so is Y (see, e.g., [FHHMZ11, p. 550]).

However, we present the following question:

Problem 241. Assume that X is a separable dual space and Y is Lipschitz homeomorphic to X. Is Y isomorphic to a dual space?

We refer to [BenLin00, p. 183].

The following few problems are special cases of Problem 239.

Problem 242. Is every Banach space that is Lipschitz homeomorphic to $C[0, 1]$ necessarily linearly isomorphic to it? A similar question for $C(K)$, where K is a countable compact (unless $C(K)$ is assumed to be linearly isomorphic to c_0).

We refer to [BenLin00] and [GoLaZi14].

Problem 243. If X is Lipschitz homeomorphic to ℓ_1, is X linearly isomorphic to ℓ_1?

We refer to [BenLin00, FHHMZ11, GoLaZi14], and [Kal08, p. 28]. *It is known to be so if X is assumed to be a dual space* [Kal08]. Let us note in passing that a similar problem to Problem 243 for c_0 has a positive solution (see, e.g., [BenLin00, p. 236]).

There are two nonseparable WCG spaces that are Lipschitz homeomorphic but not linearly isomorphic. For references, see [GoKal03, p. 133]. The density here is $\aleph_{\omega_0}$. This suggests the following:

Problem 244. Do there exist two WCG space of density ω_1 that are Lipschitz homeomorphic but not linearly isomorphic?

Problem 245. Assume that X is a Banach space Lipschitz homeomorphic to ℓ_∞. Is X linearly isomorphic to ℓ_∞?

We refer to [Kal11] and [GoLaZi14].

Problem 246. Assume that a separable Banach space X does not contain a copy of ℓ_1 and Y is Lipschitz homeomorphic to X. Is it true that Y does not contain a copy of ℓ_1?

We refer to [BenLin00] and [GoLaZi14].

Problem 247. Assume that X does not contain a copy of c_0 and Y is Lipschitz homeomorphic to X. Is it true that Y does not contain an isomorphic copy of c_0?

We refer to [BenLin00] and [GoLaZi14].

Problem 248. Let X be a reflexive Banach space of density c. Does X Lipschitz embed into ℓ_∞?

We refer to [Kal11] and [GoLaZi14]. N. J. Kalton showed in [Kal11] that *if X is of density character c and X has an unconditional basis, then X is Lipschitz embeddable in ℓ_∞.* This gives that X admits a countable family of Lipschitz real-valued functions on X that separates the points of X. Note that in $c_0(\Gamma)$, where card $\Gamma = c$, for example, this cannot happen for linear continuous functions. It is natural to ask then whether the result of N. J. Kalton mentioned in this paragraph can be used in some other geometric questions on such spaces X.

Problem 249. If c_0 Lipschitz embeds into a Banach space X, is c_0 isomorphic to a subspace of X?

We refer to [Kal08].

We mention here two papers by J. Pelant and his collaborators: [Pe94] and [PHK06]. In the second one it is shown that *there is no uniform homeomorphism of $C[0, \omega_1]$ onto a subset of any $c_0(\Gamma)$.* We do not know if this can happen for reflexive spaces. More precisely, we pose the following question:

Problem 250. Does there exist a nonseparable reflexive space that is not uniformly homeomorphic to a subset of any $c_0(\Gamma)$?

In the direction of Problem 250 let us mention that we even do not know the answer to the following problem:

Problem 251. Can $\ell_2(c)$ be Lipschitz homeomorphically mapped onto a subset of $c_0(c)$?

By $\ell_2(c)$ we mean $\ell_2(\Gamma)$, where card$(\Gamma) = c$. The same applies to $c_0(c)$. We took this problem from [GoLaZi14].

Problem 252. Assume that X admits a Fréchet differentiable norm, has density c and has an unconditional basis. Does there exist a Lipschitz homeomorphism ϕ from X onto a subset of ℓ_∞ such that $e_j^* \circ \phi$ is Gâteaux differentiable on X for each coordinate functional e_j^*?

As the Johnson–Lindenstrauss space JL_0 shows, $c_0(c)$ *is Lipschitz homeomorphic to a subspace of* ℓ_∞ (see, e.g., [FHHMZ11, p. 640]). On the other hand, we do not know the answer to the following problem:

Problem 253. Is ℓ_∞ Lipschitz homeomorphic to a subset of $c_0(c)$?

We took this problem from [GoLaZi14].

The following question is also related to Problem 239.

Problem 254 (V. Ferenczi). Let X be a separable infinite-dimensional Banach space that is not linearly isomorphic to a Hilbert space. Does there exist an infinite-dimensional subspace of X that is not Lipschitz homeomorphic to X?

We refer to [Fer03] for more on this problem.

The following is a "local" problem in this area.

Problem 255. Can one produce a short direct renorming proof of the known fact that if X is uniformly convex and Y is Lipschitz homeomorphic to X then Y can be renormed by a uniformly convex norm?

Can one use here a modification of the method of G. Godefroy, N. J. Kalton, and G. Lancien described in [FHHMZ11, p. 553]?

Connected to Problem 239 is the following open question [BenLin00, Problem 7.1]:

Problem 256. Assume that φ is a Lipschitz homeomorphism from a separable superreflexive Banach space X onto a superreflexive Banach space Y. Does there exist a point $x_0 \in X$ such that φ is Gâteaux differentiable at x_0 and the Gâteaux differentiability operator D_φ is onto Y?

In this connection let us mention that D. J. Ives and D. Preiss showed (see, e.g., [BenLin00, p. 182]) that *there is a Lipschitz homeomorphism φ from ℓ_2 onto ℓ_2 such that φ is Gâteaux differentiable at 0 and the Gâteaux differentiability operator $D_\varphi(0)$ is $-S$, where S is the right shift operator $S(a_1, a_2, \ldots) = (0, a_1, a_2, \ldots)$.*

In the direction of Problem 256 we can ask

Problem 257. Assume that φ is a Lipschitz homeomorphism of a nonseparable reflexive (superreflexive) space onto a nonseparable reflexive (respectively, superreflexive) space Y. Does there exist a point $x_0 \in X$ such that φ is Gâteaux differentiable at x_0?

We refer to [Ziz03, GoLaZi14, BenLin00] for more on this question.

Problem 258. Assume that X and Y are Lipschitz homeomorphic Banach spaces and assume that X admits an equivalent k-times Fréchet differentiable norm, for some $k \in \mathbb{N}$. Does Y admit such a norm?

The answer is known to be yes in case X is separable and k = 1. We refer to [BenLin00, p. 274] or [FHHMZ11, p. 571].

Problem 259. Assume that X and Y are Lipschitz homeomorphic Banach spaces and assume that X admits an equivalent LUR norm. Does Y admit such a norm?

We refer to [MOTV09, p. 121].

Problem 260. Assume that X and Y are Lipschitz homeomorphic Banach spaces and assume that X admits an equivalent WUR norm. Does Y admit such a norm?

The answer is known to be yes for separable spaces. Indeed, if X is separable, it admits a WUR equivalent norm if and only if X^* is separable; it is enough now to recall that a Lipschitz homeomorphism preserves this property (see, e.g., [FHHMZ11, p. 571]). *The space JL_0 of Johnson and Lindenstrauss* (see, e.g., [FHHMZ11, p. 640]) *is an example of an Asplund space that does not admit any equivalent WUR norm.*

The space ℓ_∞ can be endowed with an equivalent (dual) strictly convex norm (see, e.g., [FHHMZ11, Problem 8.62]). Consider the following question:

Problem 261. Assume that a Banach space X Lipschitz embeds into ℓ_∞. Does X admit a strictly convex norm?

We refer to [GoLaZi14].

5.3 Lipschitz Quotients

Many open problems are in the recent area of Lipschitz quotient maps. If X and Y are Banach spaces, a map f from X onto Y is called a **Lipschitz quotient map** if f is Lipschitz and there is a constant $C > 0$ such that for every $x \in X$ and for every $\varepsilon > 0$, $f(B_\varepsilon(x)) \supset B_{\varepsilon/C}(f(x))$.

Note that Lipschitz homeomorphisms, linear quotient maps, and their compositions form Lipschitz quotient maps. Note, too, that the map $f(re^{i\phi}) := re^{2i\phi}$, which is not such a composition (see [BenLin00, p. 269]), is a Lipschitz quotient map from $\mathbb{R}^2$ onto $\mathbb{R}^2$.

As we saw above, I. Aharoni and J. Lindenstrauss proved that *there is a nonseparable subspace JL_0 of ℓ_∞ that is Lipschitz homeomorphic to $c_0(\Gamma)$. The space JL_0 is thus a Lipschitz quotient of $c_0(\Gamma)$ that is not a linear quotient of $c_0(\Gamma)$*, as all WCG subspaces of ℓ_∞ are separable (see, e.g., [FHHMZ11, p. 163]).

It was proved by W. B. Johnson, J. Lindenstrauss, D. Preiss, and G. Schechtman in [JLPS02] that *there is a Lipschitz quotient map from $C[0, 1]$ onto any separable Banach space*, and that *there is a Lipschitz quotient map from ℓ_∞ onto c_0*. Note that, for example, ℓ_1 is not a linear quotient of $C[0, 1]$ and c_0 is not a linear quotient of ℓ_∞ (see, e.g., [FHHMZ11, p. 152 and p. 412]). On the other hand, *if there is a Lipschitz quotient map of a Hilbert space onto a Banach space X, then X is linearly a quotient of a Hilbert space, i.e., isomorphic to a Hilbert space* [BenLin00, p. 274].

We refer also to [BenLin00, p. 268] for more on the concept of Lipschitz quotient maps.

In the paper [BJLPS99] it is proved that *any Lipschitz quotient of L_p, $p > 1$, is a linear quotient of L_p*. However the following problem is posted there:

Problem 262. Let $1 < p < \infty$, $p \neq 2$. Is every Lipschitz quotient of ℓ_p isomorphic to a linear quotient of ℓ_p?

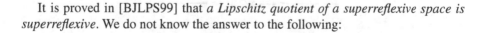

It is proved in [BJLPS99] that *a Lipschitz quotient of a superreflexive space is superreflexive*. We do not know the answer to the following:

Problem 263. Assume that X is a separable reflexive Banach space. Let Y be a Lipschitz quotient of X. Is Y reflexive?

It is known that a Lipschitz quotient of an Asplund space is an Asplund space (see, e.g., [BenLin00, p. 275]). For the linear case see, e.g., [Fa97, p. 7].

Problem 264. Let X admit a Fréchet differentiable norm and Y be a Lipschitz quotient of X. Does Y admit a Fréchet differentiable bump?

Note in passing that it may not admit a Fréchet differentiable norm (see [Hay99]).

Problem 265. Is ℓ_1 a Lipschitz quotient of the James' tree space JT?

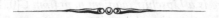

Problem 266. Let X be a weak Asplund space and Y be a Lipschitz quotient of X. Is Y a weak Asplund space?

For the linear case the answer is yes, see, e.g., [Fa97, p. 41].

The following is a natural question which arises along with Problem 32.

Problem 267. If X is a nonseparable Banach space, does there exist an infinite-dimensional separable Lipschitz quotient of X?

5.4 General Nonlinear Homeomorphisms

Problem 268. Is every quasi-Banach space homeomorphic to a Hilbert space?

We refer to [Kal03, p. 1127]. R. Cauty has shown that *there are examples of separable complete metric linear spaces that are not homeomorphic to a Hilbert space* [Cau94].

Problem 269. Is every infinite-dimensional convex compact set in a complete metric linear space homeomorphic to the **Hilbert cube**, i.e., to $[0, 1]^{\omega_0}$ in its pointwise topology?

This is the case for Banach spaces by Keller's theorem (see, e.g., [FHHMZ11, p. 541]). We refer to [Kal08, p. 14].

While *no n-dimensional Banach space can be homeomorphic to any m-dimensional Banach space if $n \neq m$* ([Dug67, p. 359], see also [FHHMZ11, p. 542]), it was already mentioned in the introduction to this Chap. 5 that *all separable infinite-dimensional Banach spaces are homeomorphic to ℓ_2* (M. I. Kadets [Ka67], see, e.g., [FHHMZ11, p. 543]), and that this was extended by H. Torunczyk [Tor81] to read: *All Banach spaces are homeomorphic to Hilbert spaces*. A "Banach space" proof is available for reflexive spaces (C. Bessaga [Be72]) and for $c_0(\Gamma)$ (S. L. Troyanski [Tro67]).

It would be interesting to have a similar "Banach space" proof of this Torunczyk's result, at least for, say, WCG spaces. More precisely, we can state the following question:

Problem 270. Can Bessaga's or Troyanski's proof be extended to WCG spaces and made the homeomorphic map coordinatewise Gâteaux smooth? What about the same smoothness question even for separable spaces?

Problem 271. Let X be an LUR renormable Banach space, and let Y be a Banach space such that in its weak topology is homeomorphic to X in its weak topology. Is Y LUR renormable?

We refer to [MOTV09, p. 121].

Not much is known on the uniform homeomorphisms of ball to spheres in infinite dimensions.

Problem 272. For which infinite-dimensional Banach spaces X is the unit sphere of X uniformly homeomorphic to the ball of X?

Actually, we do not know of any Banach space which would have this property. We refer to [BenLin00, p. 206]. *For the case of homeomorphisms in an infinite-dimensional situation, the answer is positive* [BePe75, p. 190], *unlike the situation in finite dimensions* (this is Brouwer's theorem, see, e.g., [FHHMZ11, p. 542]). *For the Lipschitz case, there are counterexamples*, see [BenLin00, p. 207].

Problem 273. Is the sphere of a superreflexive Banach space uniformly homeomorphic to the sphere in a Hilbert space?

We refer to [BenLin00, p. 218].

For a Banach space X let $\tilde{X}$ denote the space $(X \oplus \mathbb{R})_\infty$.

Problem 274. Is it true that for all infinite-dimensional Banach spaces X, $S_{\tilde{X}}$ is uniformly homeomorphic to S_X?

We refer to [BenLin00, p. 207]. In this direction, two things should be mentioned here: First, *if B_X is the unit ball of an infinite-dimensional Banach space X, then there is a Lipschitz map F from B_X into itself and $\delta > 0$ such that $\|F(x) - x\| > \delta$ for all $x \in B_X$* (P. K. Lin and Y. Sternfeld, see, e.g., [BenLin00, p. 63]). Second, *if X is an infinite-dimensional Banach space, then S_X is a Lipschitz retract of B_X* (Y. Benyamini and Y. Sternfeld, see, e.g., [BenLin00, p. 64]).

Problem 275 (N. J. Kalton). If X is a subspace of L_1, is B_X uniformly homeomorphic to B_{ℓ_2}?

We refer to [Kal04].

A **paving** of a separable Banach space X is a sequence of finite-dimensional subspaces $E_1 \subset E_2 \subset \ldots$ such that $\bigcup_{i=1}^{\infty} E_i$ is dense in X.

We will say that Banach spaces X **and** Y **have a common paving** if there are pavings $\{E_n\}_{n=1}^{\infty}$ of X and $\{F_n\}_{n=1}^{\infty}$ of Y such that the isomorphism constants of E_n and F_n stay uniformly bounded, i.e., for each n, there is an isomorphism T_n from E_n onto F_n so that $\{\|T_n\|.\|T_n^{-1}\|\}_{n=1}^{\infty}$ is a bounded sequence of numbers.

M. Ribe proved that *there are two uniformly homeomorphic Banach spaces that are not isomorphic* (see, e.g., [BenLin00, p. 246]). M. Ribe proved (see, e.g., [FHHMZ11, p. 558]) that *if X and Y are uniformly homeomorphic Banach spaces, then they are crudely finitely representable in each other* (for the definition see Remarks after Problem 2). Note that from this we get that X is isomorphic to a Hilbert space if X is uniformly homeomorphic to a Hilbert space.

However the following seems to be open:

Problem 276. If X and Y are uniformly homeomorphic separable Banach spaces, do they have a common paving?

We refer to [JLS96, p. 467].

We say that a mapping f from a Banach space X into a Banach space Y is a **coarse embedding** if there exist two nonnegative functions ρ_1 and ρ_2 on $[0, \infty)$ such that $\lim\limits_{t \to +\infty} \rho_1(t) = +\infty$ and for every $x, y \in X$,

$$\rho_1(\|x - y\|) \le \|f(x) - f(y)\| \le \rho_2(\|x - y\|).$$

Such embedding is used in many applications, for example, in the group theory or metric spaces, etc. (we refer to, e.g., [GoLaZi14]). We say that a space X **coarse embeds into a space** Y if there is a coarse embedding from X into Y.

Problem 277 (G. Godefroy). Is it true that ℓ_2 coarse embeds into any infinite-dimensional Banach space?

We refer to [GoLaZi14] for more information in this direction.

The following seems to be an open problem:

Problem 278. Is every infinite-dimensional Banach space uniformly homeomorphic to its hyperplanes?

For the Lipschitz case, there are counterexamples. For related results, we refer to [BenLin00, p. 207].

Problem 279. Can a non-normable quasi-Banach space be uniformly homeomorphic to a Banach space?

We refer to [BenLin00, p. 407].

Problem 280. If X is uniformly homeomorphic to L_p, $1 \leq p < \infty, p \neq 2$, is X isomorphic to L_p?

This is Problem 7 in [Kal08].

Problem 281. Is every Banach space that is uniformly homeomorphic to c_0 necessarily linearly isomorphic to c_0?

We refer to [BenLin00, Chap. 10]. It is known that *then X^* is linearly isomorphic to ℓ_1* (see [GoKaLa01]). We also refer to [FHHMZ11, Chap. 12] and [GoLaZi14].

N. J. Kalton showed in [Kal07] that c_0 *cannot be uniformly embedded into any reflexive space.*

Problem 282. If the unit ball of a Banach space X uniformly homeomorphically embeds into a reflexive space, is X necessarily weakly sequentially complete?

This is Problem 16 in [Kal08].

5.5 Some More Topological Problems

Problem 283 (S. Argyros, S.K. Mercourakis, and S. Negrepontis). Assume that L is a Corson compact and K is a compact space such that $C(K)$ is isomorphic to a subspace of $C(L)$. Can it be showed in ZFC that K is a Corson compact?

We took this problem from [MaPl10]. It is known that *the answer "yes" is consistent with ZFC.*

A topological space X is said to have **countable tightness** if whenever $A \subset X$ and $x \in \overline{A}$, then there is a countable set $A_0 \subset A$ such that $x \in \overline{A_0}$. The following problem is connected to the general question of the separability of the space $L_1(\mu)$, where μ is an element of $P(K)$, and K is a compact space. Here, $P(K)$ denotes the space of regular probability Radon measures on the compact space K, considered in the w^*-topology of $C(K)^*$.

Problem 284 (G. Plebanek and D. Sobota). If a compact space K is such that $P(K)$ has countable tightness, does it follow that $P(K \times K)$ has countable tightness?

We refer to [PleSo15].

A topological space T is said to be a **Baire space** if the intersection of any countable family of open dense sets in T is dense in T.

Problem 285 (M. Talagrand). Does there exist a Baire space E, a compact space K, and a real-valued function f on $E \times K$ that is nowhere continuous on $E \times K$ but everywhere separately continuous?

We recommend, e.g., [DeGoZi93, p. 328] in connection with this Talagrand's problem [Ta85], related to the so-called **Namioka property** of compact sets.

Problem 286 (S. M. Bates, A. Pełczyński). Let X be a Banach space. Does there exist a basis for the topology of X that is σ-discrete and formed by convex sets?

A family of subsets of X is called **discrete** if every point of X has a neighborhood that intersect at most one member of the family. A family of sets is σ-**discrete** if it is a union of a countable number of discrete families. A family $\mathcal{F}$ of open sets is called a **basis for the topology** if every open set is union of a subfamily of $\mathcal{F}$.

For a partial (positive) solution by M. Raja we refer to [FHHMZ11, p. 570].

The study of weak compactness in spaces of functions is a whole theory that includes many particular situations useful in several branches of functional analysis and applications. In our case, only some brief account can be given. If X and Y are Banach spaces, two main "weak" topologies on $B(X, Y)$ besides its weak topology as a Banach space have been considered: the **weak-operator topology** ω is defined by the linear functionals $T \to y^*(Tx)$, $y^* \in Y$, $x \in X$, while the **dual weak-operator topology** ω' is defined by the linear functional $T \to x^{**}(T^*y^*)$, $y^* \in Y^*$, $x^{**} \in X^{**}$. Obviously, $\omega' \succeq \omega$. If X is reflexive, then $\omega = \omega'$. Regarding the closed subspace $K(X, Y)$ of all compact operators from X into Y, a result essentially due to J. W. Brace and G. D. Friend [BrFr69] (see [Kal74]) is that *a subset A of $K(X, Y)$ is weakly compact if, and only if, A is ω'-compact*. In [Kal74], N. J. Kalton deduces a number of consequences, as: (i) *if X is reflexive, a subset A of $K(X, Y)$ is weakly compact if, and only if, it is ω-compact*, and (ii) *if X and Y are reflexive and $K(X, Y) = B(X, Y)$, then $K(X, Y)$ is reflexive* (this applies, for example, to $X := \ell_p$ and $Y := \ell_q$ if $p > q \geq 1$ by a theorem of H. R. Pitt—see, e.g., [FHHMZ11, Proposition 4.49]). *The converse to (ii) is true if X or Y has the approximation property.* Another result in [Kal74] is that *the following are equivalent: (1) X is a Grothendieck space, and (2) for any Banach space Y, if $T_n \to T$ in the weak-operator topology ω on $K(X, Y)$, then $T_n \to T$ weakly.* A **Grothendieck space** is a Banach space X such that for all separable Banach spaces Y, every bounded linear operator from X to Y is weakly compact. Extra information for the study of weak compactness in spaces of compact operators and of vector-valued functions comes from the two papers [CoRu82] and [CoRu83]: If X is a locally convex space and X'_c denote its dual space endowed with the topology of the uniform convergence on the family of all the absolutely convex compact subsets of X, the general setting treated there is the space $L_e(X'_c, Y)$, i.e., the space of continuous linear operators from X'_c into Y, endowed with the topology of the uniform convergence on the family of all equicontinuous subsets of X'. In the context of Banach spaces, they obtain again that *if X is reflexive, weak compactness in $K(X, Y)$ is equivalent to weak-operator topology compactness in the space $K(X, Y)$.* A particular consequence of their results is that, *for Banach spaces X and Y, the space $K(X, Y)$ does not contain ℓ_1 as soon as one of the two following conditions is satisfied: (a) X' does not contain ℓ_1 and Y' has the RNP, or (b) X'' has the RNP and Y does not contain ℓ_1.* The first part of the following problem asks maybe a too general question. This is why we specialized it in its second part. It is motivated by the kind of problems treated by J. Bonet and M. Friz in [BoFr02]:

Problem 287. (i) Characterize the weakly compact subsets of $B(X, Y)$, where X and Y are Banach spaces.
(ii) The same question for the case $X := \ell_1$, i.e., characterize the weakly compact subsets of the space $\ell_\infty(Y)$, where Y is a Banach space.

Chapter 6
Some More Nonseparable Problems

6.1 Schauder Basis Sets

The following is the definition of a type of Schauder basis that works also for nonseparable spaces. It is due to P. Enflo and H. P. Rosenthal in [EnRo73].

Let X be a Banach space and let S be a set in X. We say that S is a **Schauder basis set for** X provided that

 (i) the closed linear hull of S equals to X, and
(ii) every countable infinite subset of S has an ordering under which is a **Schauder basic sequence**.

We say that X has a **Schauder basis set** if X contains a set S with properties (i) and (ii) above.

It follows that a separable Banach space admits a Schauder basis set if and only if it has a Schauder basis.

P. Enflo and H. P. Rosenthal proved [EnRo73] that *if $1 < p < \infty$ and μ is a probability measure, then the space $L^p(\mu)$ has a Schauder basis set*, and that *if the density of $L^p(\mu)$ is large, then $L^p(\mu)$ does not have any unconditional Schauder basis*, unlike the separable situation (the Marcinkiewicz–Paley theorem, see, e.g., [AlKal06, p. 130]).

In [EnRo73], the authors posed the following problem:

Problem 288 (P. Enflo and H. P. Rosenthal). Let μ be a probability measure. Does $L^1(\mu)$ always have a Schauder basis set?

We also may ask

Problem 289. Study the Enflo–Rosenthal notion of a Schauder basis set in general Banach spaces.

6.2   Support Sets

Parts of nonseparable Banach space geometry are closely connected with set-theoretic concepts. Problems are sometimes even not decidable in the sole ZFC theory, i.e., under some additional consistent axioms the answer is positive, and under some other additional consistent axioms it is negative.

This is transparent, for example, with a classical question of S. Rolewicz formulated below.

Call a closed convex set C in a Banach space a **support set** if every point $x_0 \in C$ is a **proper support point**, meaning that there is an $f \in X^*$ such that

$$f(x_0) = \inf\{f(x) : x \in C\} < \sup\{f(x) : x \in C\}.$$

Using the barycenters of dense sets, one can show that *there cannot exist any support set in a separable Banach space* (S. Rolewicz, see [FHHMZ11, Lemma 12.26]), while it is not difficult to show that *any uncountable biorthogonal system generates a support set* (for details see, e.g., [HMVZ08, p. 274]). Thus, many nonseparable spaces (for example, nonseparable WCG spaces) do contain support sets (see, e.g., [HMVZ08, p. 213]).

The question stated by S. Rolewicz was to know whether every nonseparable Banach space contains a support set. This problem is not decidable in ZFC. Indeed, *under Martin (MM) axiom*, S. Todorčević showed that *all nonseparable Banach spaces do contain a support set*, while P. Koszmider and S. Todorčević independently showed that *under another consistent axioms with ZFC, $C(K)$-spaces of density $\aleph_1$ may not contain support sets*. We refer to [HMVZ08, p. 276] for more information.

However, the following question of S. Todorčević [To06] has a chance to hold. We mention that S. Todorčević showed in [To06] that *every $C(K)$-space of density greater than $\aleph_1$ contains a support set*.

Problem 290 (S. Todorčević). Is it true that every Banach space whose density is greater than $\aleph_1$ contains a support set?

We mention, too, that J. M. Borwein and J. Vanderwerff showed that *a Banach space X contains a support set if, and only if, it contains a system* $\{x_\alpha, x_\alpha^*\}_{1 \leq \alpha < \omega_1}$ *in* $X \times X^*$ *with* $x_\mu^*(x_\alpha) = 0$ *for all* $\alpha < \mu$, $x_\alpha^*(x_\alpha) = 1$ *and* $x_\mu^*(x_\alpha) \geq 0$ *for all* α (see [HMVZ08, p. 274]. We refer to [HMVZ08] and [DLAT10] for more in this direction.

<hr>

It is proved in [FGMZ04] that *if the density of a Banach space X is* $\aleph_1$, *then X is generated by* $\ell_p(\Gamma)$ *for some* $1 < p < \infty$ *if, and only if, X is generated by some space with modulus of smoothness of power type p.* The space X is said to be **generated by a space** Y if there exists a bounded linear operator from Y into X with dense range. The question of necessity of the density restriction in the mentioned result is there left open. Thus we pose the following:

Problem 291. Let $1 < p < \infty$. Let X be generated by some space with norm with modulus of smoothness of power type p. Is it generated by some $\ell_p(\Gamma)$?

<hr>

6.3 Equilateral Sets

A set S in a Banach space X is said to be **equilateral** if there is $\lambda > 0$ such that for every $x, y \in S$ with $x \neq y$, we have $\|x - y\| = \lambda$. *In every infinite-dimensional Banach space, there are equilateral sets of arbitrarily large finite cardinality* (P. Brass [Bra99] and B. V. Dekster [Dek00]). *However,* ℓ_1 *can be renormed not to have any infinite equilateral set* (P. Terenzi [Te87]). *Any renorming of* c_0 *does contain an infinite equilateral set* (S. K. Mercourakis and G. Vassiliadis [MerVa14]). *Any infinite-dimensional Banach space can be renormed to contain an infinite equilateral set* (S. K. Mercourakis and G. Vassiliadis [MerVa14]). For all of this we refer to the aforementioned [MerVa14].

It was proved in [ElOd81] that *the unit sphere of every infinite-dimensional normed space contains a* $(1 + \varepsilon)$-*separated (infinite) sequence for some* $0 < \varepsilon < 1$ (a set $S \subset X$ is said to be ε-**separated** if $\|x - y\| \geq \varepsilon$ for all $x, y \in S$, $x \neq y$).

If X is a Banach space, put

$$P(X) := \sup\{\lambda : B_X \text{ contains an infinite } \lambda\text{-separated set}\}.$$

Then it is shown in [Ko75] that $P(c_0) = 2$ and $P(\ell_p) = 2^{\frac{1}{p}}$ if $1 \leq p < \infty$.

In this notation, we pose the following problem:

Problem 292. If T is the Tsirelson space, estimate $P(T)$.

Problem 293 (P. Koszmider). Does it exist in ZFC a nonseparable Banach space with no uncountable equilateral set?

P. Koszmider proved in [Kosz] that *it is consistent that there is a nonseparable $C(K)$-space with no uncountable equilateral set* and that *it is also consistent that all nonseparable $C(K)$-spaces have uncountable equilateral sets*. D. Freeman, E. Odell, B. Sari, and Th. Schlumprecht showed in [FOSS14] that *every infinite-dimensional uniformly smooth Banach space contains an infinite equilateral set*.

Concerning Auerbach bases, B. V. Godun, B. L. Lin, and S. L. Troyanski proved that in particular, $\ell_1(c)$ *admits an equivalent norm in which it has no Auerbach basis* (see, e.g., [HMVZ08, p. 158]). We can pose the following problem:

Problem 294. Does there exist a nonseparable Banach space X with unconditional basis such that no nonseparable subspace of X has an Auerbach basis?

Chapter 7
Some Applications

7.1 Fixed Points

Schauder's theorem asserts (see the comments to Problem 274 above) that *if C is a compact convex set in a Banach space X and if f is a continuous mapping from C into C, then f has a **fixed point**,* i.e., there is $x \in C$ such that $f(x) = x$ (cf., e.g., [FHHMZ11, p. 542]).

On the other hand, V. Klee proved that *for every non-compact closed convex subset C of a Banach space there exists a continuous mapping $f : C \to C$ without a fixed point.* As we already mentioned, later, P. K. Lin and Y. Sternfeld proved that f can even be taken to be Lipschitz. More precisely, *for every non-compact closed and convex subset of a Banach space there exists a Lipschitz mapping f from C into C and there is $\delta > 0$ such that $\|f(x) - x\| \geq \delta$ for every $x \in C$* (for both results see, e.g., [BenLin00, pp. 62–63]).

If C is a bounded convex set in a Banach space X, its **Chebyshev radius** $r(C)$ is defined by $r(C) = \inf_{x \in C} \sup_{y \in C} \|x - y\|$. If for any nonsingleton (nonsingleton weakly compact) convex bounded set C in X we have that

$$\frac{r(C)}{\text{diam } C} < 1,$$

we say that X has **normal structure** (respectively, **weak normal structure**).
If

$$\sup \left\{ \frac{r(C)}{\text{diam } C} : \text{nonsingleton bounded convex } C \subset B_X \right\} < 1,$$

we say that X has **uniform normal structure**.
These concepts are used, for example, in Fixed Point Theory.

© Springer International Publishing Switzerland 2016
A.J. Guirao et al., *Open Problems in the Geometry and Analysis of Banach Spaces*,
DOI 10.1007/978-3-319-33572-8_7

Problem 295. Assume that a Banach space X has uniform normal structure. Is X necessarily superreflexive?

We refer to [BenLin00, Chap. 3] and [BenLin00, pp. 414–415] for more on this subject.

<hr/>

A mapping f from a set A in a Banach space X into a Banach space Y is called a **nonexpansive mapping** if

$$\|f(x) - f(y)\| \leq \|x - y\|$$

for all $x, y \in A$.

A Banach space X is said to satisfy the **(weak) Fixed Point Property for nonexpansive mappings ((w-)FPP**, in short) if for every convex closed bounded (respectively, for every convex weakly compact) subset C of X, every nonexpansive mapping $f : C \to C$ has a fixed point. Of course, in the class of reflexive Banach spaces the two properties coincide.

Hilbert spaces (F. E. Browder [Brow65]) and, more generally, uniformly convex Banach spaces (F. E. Browder [Brow65b]) enjoy the FPP, while the classical nonreflexive sequence spaces c_0, ℓ_1, and ℓ_∞ fail this property. An important result is that *every Banach space with weak normal structure has the w-FPP* [Ki65]. In particular, *a reflexive Banach space with normal structure has the FPP*.

The space c_0 fails to have weak normal structure (use $\overline{\mathrm{conv}}\{e_n : n \in \mathbb{N}\}$, where $\{e_n\}_{n=1}^\infty$ is the canonical basis of c_0). However, *it has the w-FPP* (B. Maurey [Mau81]). Indeed, a kind of converse holds: *a closed convex and bounded subset C of c_0 has the property that every nonexpansive mapping $f : C \to C$ has a fixed point if, and only if, C is weakly compact* [DLT04]. *The space $L_1[0, 1]$, on the contrary, fails the w-FPP* [Als81], hence both ℓ_∞ and $C[0, 1]$, being universal for the separable Banach spaces, fail the w-FPP, too.

An important and well-known problem in this area is the following:

Problem 296. Does every reflexive Banach space have the FPP?

We refer to [BenLin00, p. 80]. This problem is open even for superreflexive Banach spaces (see, e.g., [DLT96]). For a while the converse question—whether every space with the FPP should be reflexive—was also open. A counterexample was provided by the result of Pei-Kee Lin mentioned in Remarks to Problem 299.

<hr/>

The FPP and the w-FPP are metric properties, in the sense that in general they are not stable under renormings (compare the fact that $(\ell_1, \|\cdot\|_1)$ has not the FPP with the aforementioned result of Pei-Kee Lin). Problems 297–300 below address this kind of questions.

Problem 297. Let $(X, \|\cdot\|)$ be a superreflexive (or only reflexive) Banach space, let $\||\cdot\||$ be an equivalent norm on X, and B be the closed unit ball of X in the norm $\||\cdot\||$. Finally, let f be a nonexpansive mapping (in the norm $\|\cdot\|$) from the ball B into itself. Does f have a fixed point?

We refer to [FHHMZ11, p. 560] for more on this problem. Compare with the result of T. Domínguez-Benavides mentioned after Problem 298.

The following question is apparently still open, so we can repeat it from [DeGoZi93, p. 334].

Problem 298. If X is a reflexive Banach space, does X admit an equivalent norm with normal structure?

*Every separable Banach space can be renormed to be **uniformly convex in every direction**,* a property that implies normal structure [DJS71, Ziz71]. The question in Problem 298 is related to the result that *every Banach space which can be embedded in $c_0(\Gamma)$ (for instance, reflexive spaces or, more generally, spaces with an M-basis) has an equivalent renorming which enjoys the w-FPP* [DB09]. As a consequence, *every reflexive Banach can be renormed to satisfy the FPP*. Furthermore, *this norm can be chosen arbitrarily close to the original norm*.

There are some other spaces which cannot be renormed to enjoy the w-FPP, as $\ell_\infty(\Gamma)$ whenever Γ is uncountable (because *any renorming of this space contains an isometric copy of ℓ_∞* [Par81]).

Problem 299 (W. A. Kirk, Pei-Kee Lin). Does there exist an equivalent norm $\||\cdot\||$ on c_0 such that $(c_0, \||\cdot\||)$ has the FPP?

Pei-Kee Lin [Li08] proved that *there is an equivalent norm* $\vert\vert\vert \cdot \vert\vert\vert$ *on* ℓ_1 *such that* $(\ell_1, \vert\vert\vert \cdot \vert\vert\vert)$ *has the FPP*, giving the first example of a nonreflexive Banach space with FPP. We recall here the note preceding Problem 296 regarding the weak compactness of closed bounded and convex subsets of c_0 for which every nonexpansive map from C into itself has a fixed point [DLT04]. We refer also to [Ki95] for the aforementioned problem. *The space* ℓ_∞ *cannot be renormed to enjoy the FPP* [DLT01].

Problem 300 (T. Domínguez-Benavides). Can reflexive spaces be renormed to satisfy the Stable Fixed Point Property?

A Banach space X is said to satisfy the **Stable Fixed Point Property** (**SFPP**, in short) if there exists a number $\lambda > 1$ such that if Y is isomorphic to X and the Banach–Mazur distance $d(X, Y)$ is less than λ, then Y satisfies the FPP. It is known [DB12] that *any reflexive separable Banach space can be renormed to satisfy the SFPP* (in fact, for any $\lambda < (\sqrt{33} - 3)/2$). On the other hand, every space whose norm is non-distortable and fails the FPP, as ℓ_∞, cannot be renormed to satisfy the SFPP.

If X is a Banach space, then a map f from B_X into itself is called an **involution** if f^2 is the identity on B_X.

Problem 301. Let X be a Banach space and f be a uniformly continuous involution of B_X. Does there exist a sequence $\{x_n\}_{n=1}^\infty$ of points of B_X such that $\|f(x_n) - x_n\| \to 0$?

We took this problem from [BenLin00, p. 217]. As stated there, *if B_X were uniformly homeomorphic to S_X, then we would get a counterexample here.*

7.2   Riemann Integrability of Vector-Valued Functions

A bounded function $f : [0, 1] \to \mathbb{R}$ is Riemann integrable if, and only if, it is almost everywhere continuous on $[0, 1]$, according to a classical characterization of B. Riemann and H. Lebesgue (see, e.g., [MZZ15]). If f has values in a general Banach space X, the result is no longer true. The definition of being Riemann integrable is done similarly to the real case: A function $f : [0, 1] \to X$ is **Riemann integrable** if there exists a vector $x \in X$ (its **Riemann integral** $\int_0^1 f(t)\,dt$) such that, for each $\varepsilon > 0$ there exists a partition P_ε of $[0, 1]$ such that for every refinement P of P_ε and tags t_i (one in each subinterval S_i of P), we have $| \sum f(t_i)\lambda(S_i) - x| < \varepsilon$, where $\lambda(S_i)$ is the length of S_i. A Banach space X has the **Lebesgue property** if every Riemann integrable function $f : [0, 1] \to X$ is continuous almost everywhere on $[0, 1]$ (the reverse implication is always true). The spaces c, c_0, $C[a, b]$, ℓ_p for $1 < p \leq \infty$, $L_1[a, b]$, $L_\infty[a, b]$, the dual space X^* if X contains a copy of ℓ_1, and every infinite-dimensional uniformly convex Banach space fail the Lebesgue property. The space ℓ_1 has the Lebesgue property, and so they do the so-called **asymptotic ℓ_1 spaces**. For the definition and the former results see, e.g., [Gor91]. The following problem remains, to our best knowledge, open.

Problem 302. Characterize the Banach spaces having the Lebesgue property.

We refer to, e.g., [MZZ15] for basics on Riemann integration.

7.3   Miscellaneous

Problem 303 (L. Nirenberg). Assume that f is a continuous map from a Hilbert space H into itself. Assume that there is a constant $k > 0$ such that $\|f(x) - f(y)\| \geq k\|x - y\|$ for all $x, y \in H$. Assume further that the image $f(H)$ has a nonempty interior. Is f necessarily onto?

We refer to [BenLin00, p. 217].

M. Talagrand proved that *the space $C[0, \omega_1]$ does not admit any norm whose dual norm would be strictly convex* [Ta85], see also [DeGoZi93, p. 313].

Problem 304. Can one use some version of the variational principle to reprove Talagrand's result mentioned above by showing that the supremum norm on $C[0, \omega_1]$ is a kind of "too rough" to live together with a norm whose dual is strictly convex?

References

[Aco99] M.D. Acosta, Denseness of norm attaining operators into strictly convex spaces. Proc. R. Soc. Edinb. A **129**, 1107–1114 (1999)

[Aco06] M.D. Acosta, Denseness of norm attaining mappings. Rev. R. Acad. Cienc. Exactas Fís. Nat. Ser. A Math. RACSAM **100**(1–2), 9–30 (2006)

[AcAgPa96] A.D. Acosta, F.J. Aguirre, R. Payá, There is no bilinear Bishop-Phelps theorem. Isr. J. Math. **93**, 221–227 (1996)

[AAAG07] M.D. Acosta, A. Aizpuru, R.M. Aron, F.J. García-Pacheco, Functionals that do not attains their norm. Bull. Belg. Math. Soc. Simon Stevin **14**(3), 407–418 (2007)

[AcoKa11] M.D. Acosta, V. Kadets, A characterization of reflexive spaces. Math. Ann. **349**, 577–588 (2011)

[AcoMon07] M.D. Acosta, V. Montesinos, On a problem of Namioka on norm-attaining functionals. Math. Z. **256**, 295–300 (2007)

[AhaLi78] I. Aharoni, J. Lindenstrauss, Uniform equivalence between Banach spaces. Bull. Am. Math. Soc. **84**, 281–283 (1978)

[ACPP05] G. Alberti, M. Csörnyei, A. Pełczyński, D. Preiss, BV has the bounded approximation property. J. Geom. Anal. **15**(1), 1–7 (2005)

[AlKal06] F. Albiac, N.J. Kalton, *Topics in Banach Space Theory*. Graduate Texts in Mathematics, vol. 233 (Springer, Berlin, 2006)

[AlKal09] F. Albiac, N.J. Kalton, Lipschitz structure of quasi-Banach spaces. Isr. J. Math. **170**, 317–335 (2009)

[Als81] D.E. Alspach, A fixed point free nonexpansive mapping. Proc. Am. Math. Soc. **82**, 423–424 (1981)

[AlsSa16] D.E. Alspach, B. Sari, Separable elastic spaces are universal. J. Funct. Anal. **270**(1), 177–200 (2016)

[AmLin68] D. Amir, J. Lindenstrauss, The structure of weakly compact sets in Banach spaces. Ann. Math. **88**, 35–46 (1968)

[And72] R.D. Anderson, Hilbert space is homeomorphic to the countable product of lines. Bull. Am. Math. Soc. **72**, 515–519 (1966)

[ACK00] G. Androulakis, P.G. Casazza, D. Kutzarova, Some more weak Hilbert spaces. Can. Math. Bull. **43**, 257–267 (2000)

[An10] R. Anisca, The Ergodicity of weak Hilbert spaces. Proc. Am. Math. Soc. **208**(4), 1405–1413 (2010)

[Ans97] S. Ansari, Existence of hypercyclic operators on topological vector spaces. J. Funct. Anal. **148**, 384–390 (1997)

[ArgArv04] S.A. Argyros, A.D. Arvanitakis, Regular averaging and regular extension operators in weakly compact subsets of Hilbert spaces. Serdica Math. J. **30**, 527–548 (2004)

[ABR12] S.A. Argyros, K. Beanland, T. Raikoftsalis, An extremely non-homogeneous weak Hilbert space. Trans. Am. Math. Soc. **364**(9), 4987–5014 (2012)

[ACGJM02] S.A. Argyros, J.F. Castillo, A.S. Granero, M. Jiménez, J. Moreno, Complementation and embeddings of $c_0(I)$ in Banach spaces. Proc. Lond. Math. Soc. **85**(3), 742–768 (2002)

[ArDoKa08] S.A. Argyros, P. Dodos, V. Kanellopoulos, A class of separable Rosenthal compacta and its application. Diss. Math. **449**, 1–52 (2008)

[ArGoRo03] S.A. Argyros, G. Godefroy, H.P. Rosenthal, *Descriptive Set Theory and Banach Spaces*, ed. by W.B. Johnson, J. Lindenstrauss. Handbook of the Geometry of Banach Spaces II (Elsevier, Amsterdam, 2003), pp. 1007–1069

[ArHay11] S.A. Argyros, R.G. Haydon, A hereditarily indecomposable $\mathcal{L}_\infty$-space that solves the scalar-plus-compact problem. Acta Math. **206**, 1–54 (2011)

[ArMe93] S.A. Argyros, S.K. Mercourakis, On weakly Lindelöf Banach spaces. Rocky Mt. J. Math. **23**, 395–446 (1993)

[ArMe05] S.A. Argyros, S.K. Mercourakis, Examples concerning heredity problems of WCG Banach spaces. Proc. Am. Math. Soc. **133**(3), 773–785 (2005)

[AJR13] J.M. Aron, J.A. Jaramillo, T. Ransford, Smooth surjections without surjective restrictions. J. Geom. Anal. **23**, 2081–2090 (2013)

[Arsz76] N. Aronszajn, Differentiability of Lipschitzian mappings between Banach spaces. Stud. Math. **57**, 147–190 (1976)

[As68] E. Asplund, Fréchet differentiability of convex functions. Acta Math. **121**, 31–47 (1968)

[AviKa10] A. Avilés, O.F.K. Kalenda, Compactness in Banach space theory—selected problems. Rev. R. Acad. Cienc. Exactas Fís. Nat. Ser. A Math. RACSAM **104**(2), 337–352 (2010)

[AviKo13] A. Avilés, P. Koszmider, A Banach space in which every injective operator is surjective. Bull. Lond. Math. Soc. **45**, 1065–1074 (2013)

[Aza97] D. Azagra, Diffeomorphisms between spheres and hyperplanes in infinite-dimensional Banach spaces. Stud. Math. **125**(2), 179–186 (1997)

[ADJ03] D. Azagra, R. Deville, M. Jiménez-Sevilla, On the range of the derivatives of a smooth function between Banach spaces. Math. Proc. Camb. Philos. Soc. **134**(1), 163–185 (2003)

[AzaMu15] D. Azagra, C. Mudarra, Global approximation of convex functions by differentiable convex functions on Banach spaces. J. Convex Anal. **22**(4), 1197–1205 (2015)

[BaGo06] P. Bandyopadhyay, G. Godefroy, Linear structures in the set of norm-attaining functionals on a Banach space. J. Convex Anal. **13**(3–4), 489–497 (2006)

[BaHa08] M. Bačák, P. Hájek, Mazur intersection property for Asplund spaces. J. Funct. Anal. **255**, 2090–2094 (2008)

[Bat97] S.M. Bates, On smooth, nonlinear surjections of Banach spaces. Isr. J. Math. **100**, 209–220 (1997)

[BJLPS99] S.M. Bates, W.B. Johnson, J. Lindenstrauss, D. Preiss, G. Schechtman, Affine approximation of Lipschitz functions and nonlinear quotients. Geom. Funct. Anal. (GAFA) **9**(6), 1092–1127 (1999)

[Bay05] F. Bayart, Linearilty of sets of strange functions. Mich. Math. J. **53**, 291–303 (2005)

[BayGri06] F. Bayart, S. Grivaux, Frequently hypercyclic operators. Trans. Am. Math. Soc. **358**(11), 5083–5117 (2006)

[BayMat09] F. Bayart, É. Matheron, *Dynamics of Linear Operators*. Cambridge Tracts in Mathematics, vol. 179 (Cambridge University Press, Cambridge, 2009)

[BenLin00] Y. Benyamini, J. Lindenstrauss, *Geometric Nonlinear Functional Analysis, Vol. I*. Colloquium Publications, vol. 48 (American Mathematical Society, Providence, 2000)

[BerKal01] T. Bermúdez, N.J. Kalton, The range of operators on von Neumann algebras. Proc. Am. Math. Soc. **130**(5), 1447–1455 (2001)

[Ber99] L. Bernal-González, On hypercyclic operators on Banach spaces. Proc. Am. Math. Soc. **127**, 1003–1010 (1999)

[BesPe99] J. Bès, A. Peris, Hereditarily hypercyclic operators. J. Funct. Anal. **167**(1), 94–112 (1999)

[Be72] C. Bessaga, Topological equivalence of nonseparable reflexive Banach spaces. Ann. Math. Stud. **69**, 3–14 (1972)

[BePe75] C. Bessaga, A. Pełczyński, *Selected topics in Infinite-Dimensional Topology* (Polish Scientific Publishers, Warszawa, 1975)

[BePeRo61] C. Bessaga, A. Pelczynski, S. Rolewicz, On diametral approximative dimension and linear homogeneity of F-spaces. Bull. Acad. Polon. Sci. Sér. Sci. Math. Astronom. Phys. **9**, 677–683 (1961)

[BiSm16] V. Bible, R.R. Smith, Smooth and polyhedral approximation in Banach spaces. J. Math. Anal. Appl. **435**(2), 1262–1272 (2016)

[BiPhe61] E. Bishop, R.R. Phelps, A proof that every Banach space is subreflexive. Bull. Am. Math. Soc. **67**, 97–98 (1961)

[Boh41] F. Bohnenblust, Subspaces of $\ell_{p,n}$ spaces. Am. J. Math. **63**, 64–72 (1941)

[Bo00] J. Bonet, Hypercyclic and chaotic convolution operators. J. Lond. Math. Soc. **62**(2), 253–262 (2000)

[BoFr02] J. Bonet, M. Friz, Weakly compact composition operators on locally convex spaces. Math. Nachr. **245**, 26–44 (2002)

[BorFa93] J.M. Borwein, M. Fabian, On convex functions having points of Gâteaux differentiability which are not points of Fréchet differentiability. Can. J. Math. **45**(6), 1121–1134 (1993)

[BJM02] J.M. Borwein, M. Jiménez-Sevilla, J.P. Moreno, Antiproximinal norms in Banach spaces. J. Approx. Theory **114**, 57–69 (2002)

[BorNo94] J.M. Borwein, D. Noll, Second order differentiability of convex functions: A. D. Alexandrov's theorem in Hilbert space. Trans. Am. Math. Soc. **342**, 43–81 (1994)

[BorVan10] J.M. Borwein, J.D. Vanderwerff, *Convex Functions: Constructions, Characterizations and Counterexamples* (Cambridge University Press, Cambridge, 2010)

[BGK96] B. Bossard, G. Godefroy, R. Kaufman, Hurewicz's theorems and renormings of Banach spaces. J. Funct. Anal. **140**, 142–150 (1996)

[Bou77] J. Bourgain, On dentability and the Bishop-Phelps property. Isr. J. Math. **78**(4), 265–271 (1977)

[Bou80] J. Bourgain, Dentability and finite dimensional decompositions. Stud. Math. **67**(2), 135–148 (1980)

[BouDel81] J. Bourgain, F. Delbaen, A class of $\mathcal{L}_\infty$ spaces. Acta Math. **145**, 155–176 (1981)

[Bour83] R.D. Bourgin, *Geometric Aspects of Convex Sets with the Radon–Nikodým Property.* Lecture Notes in Mathematics, vol. 993 (Springer, Berlin, 1983)

[BrFr69] J.W. Brace, G.D. Friend, Weak convergence of sequences in function spaces. J. Funct. Anal. **4**, 457–466 (1969)

[Bra99] P. Brass, On equilateral simplices in normed spaces. Beiträge Algebra Geom. **40**, 303–307 (1999)

[Brow65] F.E. Browder, Fixed point theorems for noncompact mappings in Hilbert spaces. Proc. Natl. Acad. Sci. USA **43**, 1272–1276 (1965)

[Brow65b] F.E. Browder, Nonexpansive nonlinear operators in a Banach space. Proc. Natl. Acad. Sci. USA **54**, 1041–1044 (1965)

[Cass01] P.G. Casazza, *Approximation Properties*, ed. by W.B. Johnson, J. Lindenstrauss. Handbook of the Geometry of Banach Spaces I (Elsevier, Amsterdam, 2001), pp. 271–316

[CJT84] P.G. Casazza, W.B. Johnson, L. Tzafriri, On Tsirelson's space. Isr. J. Math. **47**(2–3), 81–98 (1984)

[CaSh89] P.G. Casazza, T. Shura, *Tsirelson's Space*. Lecture Notes in Mathematics, vol. 1363 (Springer, Berlin, 1989)

[CasGon97] J.M.F. Castillo, M. González, *Three-Space Problems in Banach Space Theory*. Lecture Notes in Mathematics, vol. 1667 (Springer, Berlin, 1997)

[Cau94] R. Cauty, Un espace métrique linéaire qui n'est pas un rétracte absolu. Fund. Math. **146**, 85–99 (1994)

[Chr72] J.P.R. Christensen, On sets of Haar measure zero in Abelian Polish groups. Isr. J. Math. **13**, 255–260 (1972)

[CoRu82] H.S. Collins, W. Ruess, Duals of spaces of compact operators. Stud. Math. **74**(3), 213–245 (1982)

[CoRu83] H.S. Collins, W. Ruess, Weak compactness in spaces of compact operators and of vector-valued functions. Pac. J. Math. **106**(1), 45–71 (1983)

[CoPe05] J.A. Conejero, A. Peris, Linear transitivity criteria. Topol. Appl. **153**, 767–773 (2005)

[CorLin66] H.H. Corson, J. Lindenstrauss, On weakly compact subsets of Banach spaces. Proc. Am. Math. Soc. **17**, 407–412 (1966)

[CDW] M. Cúth, M. Doucha, P. Wojtaszczyk, On the structure of Lipschitz-free spaces, to appear

[Dal15] A. Dalet, Free spaces over countable compact metric spaces. Proc. Am. Math. Soc. **143**(8), 3537–3546 (2015)

[DJS71] M.M. Day, R.C. James, S.L. Swaminathan, Normed linear spaces that are uniformly convex in every direction. Can. J. Math. **23**(6), 1051–1059 (1971)

[BerVe09] C.A. De Bernardi, L. Veselý, On support points and support functionals of convex sets. Isr. J. Math. **171**, 15–27 (2009)

[DRRe09] M. De la Rosa, C. Read, A hypercyclic operator whose direct sum $T \oplus T$ is not hypercyclic. J. Oper. Theory **61**(2), 369–380 (2009)

[DGS95] G. Debs, G. Godefroy, J. Saint-Raymond, Topological properties of the set of norm-attaining linear functionals. Can. J. Math. **47**(2), 318–329 (1995)

[Dek00] B.V. Dekster, Simplexes with prescribed edge lengths in Minkowski and Banach spaces. Acta Math. Hung. **86**(4), 343–358 (2000)

[De87] R. Deville, Un théorème de transfert pour la propriété des boules. Can. Math. Bull. **30**, 295–300 (1987)

[DeGh01] R. Deville, N. Ghoussoub, *Perturbed Minimization Principles and Applications*, ed. by W.B. Johnson, J. Lindenstrauss. Handbook of the Geometry of Banach Spaces I (Elsevier, Amsterdam, 2001), pp. 393–435

[DGHZ87] R. Deville, G. Godefroy, D.E.G. Hare, V. Zizler, Differentiability of convex functions and the convex point of continuity property. Isr. J. Math. **59**, 245–255 (1987)

[DGHZ88] R. Deville, G. Godefroy, D.E.G. Hare, V. Zizler, Phelps spaces and finite-dimensional decompositions. Bull. Aust. Math. Soc. **37**(2), 263–271 (1988)

[DeGoZi90] R. Deville, G. Godefroy, V. Zizler, The three space problem for smooth partitions of unity and $C(K)$ spaces. Math. Ann. **288**, 613–625 (1990)

[DeGoZi93] R. Deville, G. Godefroy, V. Zizler, *Smoothness and Renormings in Banach Spaces*. Pitman Monographs, vol. 64 (London, Longman, 1993)

[DeGoZi93b] R. Deville, G. Godefroy, V. Zizler, Smooth bump functions and the geometry of Banach spaces. Mathematika **40**, 305–321 (1993)

[DeHa05] R. Deville, P. Hájek, On the range of the derivative of Gâteaux-smooth functions on separable Banach spaces. Isr. J. Math. **145**, 257–269 (2005)

[DeIvLa15] R. Deville, I. Ivanov, S. Lajara, Construction of pathological Gâteaux differentiable functions. Proc. Am. Math. Soc. **143**(1), 129–139 (2015)

[DePro09] R. Deville, A. Procházka, A parametric variational principle and residuality. J. Funct. Anal. **256**(2), 3568–3587 (2009)

[DeZi88] R. Deville, V. Zizler, A note on cotype of smooth spaces. Manuscripta Math. **62**, 375–382 (1988)

[Di84] J. Diestel, *Sequences and Series in Banach Spaces*. Graduate Texts in Mathematics, vol. 92 (Springer, New York, 1984)

[DiUh77] J. Diestel, J.J. Uhl, *Vector Measures*. Mathematical Surveys, vol. 15 (American Mathematical Society, Providence, 1977)

[DilRan15] S.J. Dilworth, B. Randrianantoanina, On an isomorphic Banach-Mazur rotation problem and maximal norms in Banach spaces. J. Funct. Anal. **268**, 1587–1611 (2015)

[DLAT10] P. Dodos, J. López-Abad, S. Todorčević, Banach spaces and Ramsey theory: some open problems. Rev. R. Acad. Cienc. Exactas Fís. Natl. Ser. A Math. RACSAM **104**(2), 435–450 (2010)

[DB09] T. Domínguez-Benavides, A renorming of some nonseparable Banach spaces with the fixed point property. J. Math. Anal. Appl. **350**(2), 525–530 (2009)

[DB12] T. Domínguez-Benavides, Distortion and stability of the fixed point property for non-expansive mappings. Nonlinear Anal. **75**, 3229–3234 (2012)

[DLT96] P.N. Dowling, C.J. Lennard, B. Turett, Reflexivity and the fixed-point property for nonexpansive mappings. J. Math. Anal. Appl. **200**, 653–662 (1996)

[DLT01] P.N. Dowling, C.J. Lennard, B. Turett, *Renorming of ℓ_1 and c_0 and Fixed Point Properties*. Handbook of Metric Fixed Point Theory (Kluwer, Dordrecht, 2001), pp. 269–297

[DLT04] P.N. Dowling, C.J. Lennard, B. Turett, Weak compactness is equivalent to the fixed point property in c_0. Proc. Am. Math. Soc. **132**(6), 1659–1666 (2004)

[Dub71] E. Dubinsky, Every separable Fréchet space contains a non stable dense subspace. Stud. Math. **40**, 77–79 (1971)

[Dug67] J. Dugundji, *Topology* (Allyn and Bacon, Boston, 1967)

[DuFe06] Y. Dutrieux, V. Ferenczi, The Lipschitz-free Banach spaces of C(K)-spaces. Proc. Am. Math. Soc. **134**(4), 1039–1044 (2006)

[DuLan08] Y. Dutrieux, G. Lancien, Isometric embedding of compact spaces into Banach spaces. J. Funct. Anal. **255**(2), 494–501 (2008)

[Edd91] A. Eddington, Some more weak Hilbert spaces. Stud. Math. **100**(1), 1–11 (1991)

[EdTh72] M. Edelstein, A.C. Thompson, Some results on nearest points and support properties of convex sets in c_0. Pac. J. Math. **40**, 553–560 (1972)

[Edg79] G.A. Edgar, A long James space, in *Proceedings Conference Oberwolfach 1979*. Lecture Notes in Mathematics, vol. 794 (Springer, Berlin, 1979)

[ElOd81] J. Elton, E. Odell, The unit ball of every infinite-dimensional normed space contains a $(1 + \varepsilon)$ separated sequence. Colloq. Math. **44**(1), 105–109 (1981)

[EnGuSe14] P. Enflo, V. Gurarii, J.B. Seoane-Sepúlveda, Some results and open questions on spaceability in function spaces. Trans. Am. Math. Soc. **366**(2), 611–625 (2014)

[EnLiPi75] P. Enflo, J. Lindenstrauss, G. Pisier, On the three space problem. Math. Scand. **36**, 189–210 (1975)

[EnLo01] P. Enflo, V. Lomonosov, *Some Aspects of the Invariant Subspace Problem*, ed. by W.B. Johnson, J. Lindenstrauss. Handbook of the Geometry of Banach Spaces I (Elsevier, Amsterdam, 2001), pp. 533–559

[EnRo73] P. Enflo, H.P. Rosenthal, Some results concerning $L^p(\mu)$-spaces. J. Funct. Anal. **14**, 325–348 (1973)

[Fa85] M. Fabian, Lipschitz smooth points of convex functions and isomorphic characterizations of Hilbert spaces. Proc. Lond. Math. Soc. **51**, 113–126 (1985)

[Fa97] M. Fabian, *Gâteaux Differentiability of Convex Functions and Topology—Weak Asplund Spaces* (Wiley, New York, 1997)

[FaGo88] M. Fabian, G. Godefroy, The dual of every Asplund space admits a projectional resolution of idenity. Stud. Math. **91**, 141–151 (1988)

[FGMZ04] M. Fabian, G. Godefroy, V. Montesinos, V. Zizler, Inner characterizations of weakly compactly generated Banach spaces and their relatives. J. Math. Anal. Appl. **297**(2), 419–455 (1988)

[FHHMZ11] M. Fabian, P. Habala, P. Hájek, V. Montesinos, V. Zizler, *Banach Space Theory: The Basis for Linear and Nonlinear Analysis*. CMS Books in Mathematics (Canadian Mathematical Society/Springer, 2011)

[FMZ02] M. Fabian, V. Montesinos, V. Zizler, Pointwise semicontinuous smooth norms. Arch. Math. **78**, 459–464 (2002)

[FMZ06] M. Fabian, V. Montesinos, V. Zizler, Smoothness in Banach spaces. Selected problems. Rev. R. Acad. Cienc. Exactas Fís. Natl. Ser. A Math. RACSAM **100**(1–2), 101–125 (2006)

[FMZ07] M. Fabian, V. Montesinos, V. Zizler, Weak compactness and σ-Asplund generated Banach spaces. Stud. Math. **181**, 125–152 (2007)

[FPWZ89] M. Fabian, D. Preiss, J.H.M. Whitfield, V. Zizler, Separating polynomials on Banach spaces. Q. J. Math. Oxf. **40**(2), 409–422 (1989)

[Fer97] V. Ferenczi, A uniformly convex hereditarily indecomposable Banach space. Isr. J. Math. **102**, 199–225 (1997)

[Fer03] V. Ferenczi, Lipschitz homogeneous Banach spaces. Q. J. Math. **54**(4), 415–419 (2003)

[FiJo74] T. Figiel, W.B. Johnson, A uniformly convex Banach space which contains no ℓ_p. Compos. Math. **29**, 179–190 (1974)

[FiPeJo11] T. Figiel, A. Pełczyński, W.B. Johnson, Some approximation of Banach spaces and Banach lattices. Isr. J. Math. **183**, 199–231 (2011)

[FleMoo15] J. Fletcher, W.B. Moors, Chebyshev sets. J. Aust. Math. Soc. **98**, 161–231 (2015)

[FoLin98] V.P. Fonf, J. Lindenstrauss, Some results on infinite-dimensional convexity. Isr. J. Math. **108**, 13–32 (1998)

[FLP01] V.P. Fonf, J. Lindenstrauss, R.R. Phelps, *Infinite-Dimensional Convexity*, ed. by W.B. Johnson, J. Lindenstrauss. Handbook of the Geometry of Banach Spaces I (Elsevier, Amsterdam, 2001), pp. 599–670

[FPST08] V.P. Fonf, A.J. Pallarés, R.J. Smith, S.L. Troyanski, Polyhedral norms on non-separable Banach spaces. J. Funct. Anal. **255**, 449–470 (2008)

[Fos92] M. Fosgerau, A Banach space with Lipschitz Gâteaux smooth bump has weak* fragmentable dual. Ph.D Dissertation, The University College, London (1992)

[FOSS14] D. Freeman, E. Odell, B. Sari, Th. Schlumprecht, Equilateral sets in uniformly smooth Banach spaces. Mathematika **60**, 219–231 (2014)

[Fry04] R. Fry, Approximation of functions with bounded derivative on Banach spaces. Bull. Aust. Math. Soc. **69**, 125–131 (2004)

[Geo91] P.G. Georgiev, On the residuality of the set of norms having Mazur's intersection property. Math. Balkanica **5**, 20–26 (1991)

[Go87] G. Godefroy, Boundaries of a convex set and interpolation sets. Math. Ann. **277**, 173–194 (1987)

[Go00] G. Godefroy, Some applications of Simons' inequality. Serdica Math. J. **26**, 59–78 (2000)

[Go01] G. Godefroy, *Renormings of Banach Spaces* ed. by W.B. Johnson, J. Lindenstrauss. Handbook of the Geometry of Banach Spaces I (Elsevier, Amsterdam, 2001), pp. 781–835

[Go01a] G. Godefroy, The Banach space c_0. Extracta Math. **16**(1), 1–25 (2001)

[Go10] G. Godefroy, Remarks on non-linear embeddings between Banach spaces. Houst. J. Math. **36**(1), 283–287 (2010)

[Go10a] G. Godefroy, On the diameter of the Banach–Mazur set. Czechoslov. Math. J. **135**, 95–100 (2010)

[Go14] G. Godefroy, *Fréchet Differentiability of Lipschitz Functions and Porous Sets in Banach Spaces*. Annals of Mathematics Studies, vol. 179 (Princeton University Press, Princeton 2012); Review on "J. Lindenstrauss, D. Preiss, J. Tišer. Bull. Am. Math. Soc. **51**, 515–517 (2014)

[GoKal89] G. Godefroy, N.J. Kalton, The ball topology and its applications. Contemp. Math. **85**, 195–238 (1989)

[GoKal03] G. Godefroy, N.J. Kalton, Lipschitz-free Banach spaces. Stud. Math. **159**(1), 121–141 (2003)

[GoKaLa00] G. Godefroy, N.J. Kalton, G. Lancien, Subspaces of $c_0(\mathbb{N})$ and Lipschitz isomorphisms. Geom. Funct. Anal. (GAFA) **10**(4), 798–820 (2000)

[GoKaLa01] G. Godefroy, N.J. Kalton, G. Lancien, Szlenk indices and homeomorphisms. Trans. Am. Math. Soc. **353**(10), 3895–3918 (2001)

[GoLaZi14] G. Godefroy, G. Lancien, V. Zizler, The non-linear geometry of Banach spaces after Nigel Kalton. Rocky Mt. J. Math. **44**(5), 1529–1583 (2014)

[GoMZ95] G. Godefroy, V. Montesinos, V. Zizler, Strong subdifferentiability of norms and geometry of Banach spaces. Comment. Math. Univ. Carol. **36**(3), 493–502 (1995)

[GoOz14] G. Godefroy, N. Ozawa, Free Banach spaces and the approximation properties. Proc. Am. Math. Soc. **142**(5), 1681–1687 (2014)

[GoSa88] G. Godefroy, P.D. Saphar, Duality in spaces of operators and smooth norms on Banach spaces. Ill. J. Math. **32**(4), 672–695 (1988)

[GoSha98] G. Godefroy, J.H. Shapiro, Operators with dense, invariant, cyclic vector manifolds. J. Funct. Anal. **98**, 229–269 (1991)

[Gor91] R. Gordon, Riemann integration in Banach spaces. Rocky Mt. J. Math. **21**(3), 923–949 (1991)

[Gow90] W.T. Gowers, Symmetric block bases of sequences with large average growth. Isr. J. Math. **69**, 129–151 (1990)

[Gow94] W.T. Gowers, A solution to Banach's hyperplane problem. Bull. Lond. Math. Soc. **26**, 523–530 (1994)

[Gow94b] W.T. Gowers, A Banach space not containing c_0, ℓ_1, or a reflexive subspace. Trans. Am. Math. Soc. **344**, 407–420 (1994)

[Gow96] W.T. Gowers, A new dichotomy for Banach spaces. Geom. Funct. Anal. (GAFA) **6**, 1083–1093 (1996)

[GowMau93] W.T. Gowers, B. Maurey, The unconditional basic sequence problem. J. Am. Math. Soc. **6**, 851–874 (1993)

[GJM04] A.S. Granero, M. Jiménez-Sevilla, J.P. Moreno, Intersections of closed balls and geometry of Banach spaces. Extracta Math. **19**(1), 55–92 (2004)

[Gro03] K.G. Grosse-Erdmann, Recent developements in hypercyclicity. Rev. R. Acad. Cienc. Exactas Fís. Natl. Ser. A Math. RACSAM **97**, 273–286 (2003)

[GroPe11] K.G. Grosse-Erdmann, A. Peris, *Linear Chaos*. Universitext (Springer, London, 2011)

[GHM10] A.J. Guirao, P. Hájek, V. Montesinos, Ranges of operators and derivatives. J. Math. Anal. Appl. **367**, 29–33 (2010)

[GMZ12] A.J. Guirao, V. Montesinos, V. Zizler, On a classical renorming construction of V. Klee. J. Math. Anal. Appl. **385**, 458–465 (2012)

[GMZ14] A.J. Guirao, V. Montesinos, V. Zizler, *On Preserved and Unpreserved Points*, ed. by J.C. Ferrando, M.López-Pellicer. Descriptive Topology and Functional Analysis, in Honour of Jerzy Kakol's 60th Birthday. Springer Proceedings in Mathematics & Statistics, vol. 80 (Springer, Berlin, 2014), pp. 163–193

[GMZ15] A.J. Guirao, V. Montesinos, V. Zizler, A note on extreme points of C^∞-smooth balls in polyhedral spaces. Proc. Am. Math. Soc. **143**(8), 3413–3420 (2015)

[GMZ] A.J. Guirao, V. Montesinos, V. Zizler, Remarks on the set of norm-attaining functionals, to appear

[Ha77] J. Hagler, A counterexample to several questions about Banach spaces. Stud. Math. **60**, 289–308 (1977)

[Ha87] J. Hagler, A note on separable Banach spaces with nonseparable dual. Proc. Am. Math. Soc. **99**(3), 452–454 (1987)

[Haj98] P. Hájek, Smooth functions on c_0. Isr. J. Math. **104**, 17–27 (1998)

[HaJo04] P. Hájek, M. Johanis, Characterization of reflexivity by equivalent renorming. J. Funct. Anal. **211**, 163–172 (2004)

[HaJo14] P. Hájek, M. Johanis, *Smooth Analysis in Banach Spaces* (De Gruyter, Berlin, 2014)

[HLP] P. Hájek, G. Lancien, E. Pernecká, Lipschitz-free spaces over metric spaces homeomorphic to the Cantor set, to appear

[HajMon10] P. Hájek, V. Montesinos, Boundedness of biorthogonal systems in Banach spaces. Isr. J. Math. **177**, 145–154 (2010)

[HMVZ08] P. Hájek, V. Montesinos, J. Vanderwerff, V. Zizler, *Biorthogonal Systems in Banach Spaces*. CMS Books in Mathematics (Canadian Mathematical Society/Springer, 2008)

[HMZ12] P. Hájek, V. Montesinos, V. Zizler, Geometry and Gâteaux smoothness in separable Banach spaces. Oper. Matrices **6**(2), 201–232 (2012)

[HaPe14] P. Hájek, E. Pernecká, On Schauder bases in Lipschitz-free spaces. J. Math. Anal. Appl. **416**(2), 629–646 (2014)

[HajZiz06] P. Hájek, V. Zizler, Functions locally dependent on finitely many coordinates. Rev. Real Acad. Cienc. Serie A. Mat. **100**(1–2), 47–154 (2006)

[Hay90] R. Haydon, A counterexample to several questions about scattered compact spaces. Bull. Lond. Math. Soc. **22**, 261–268 (1990)

[Hay99] R. Haydon, Trees in Renorming Theory. Proc. Lond. Math. Soc. **78**(3), 541–584 (1999)

[Hay08] R. Haydon, Locally uniformly rotund norms in Banach spaces and their duals. J. Funct. Anal. **254**, 2023–2039 (2008)

[HOS11] R. Haydon, E. Odell, T. Schlumprecht, Small subspaces of L_p. Ann. Math. **173**, 169–209 (2011)

[HayZi89] R. Haydon, V. Zizler, A new space with no locally uniformly rotund renorming. Canad. Math. Bull. **32**, 122–128 (1989)

[HeMa82] S. Heinrich, P. Mankiewicz, Application of ultraproducts to uniform and Lipschitz classification of Banach spaces. Stud. Math. **73**, 225–251 (1982)

[Hu80] R.E. Huff, On non-density of norm-attaining operators. Rev. Roum. Math. Pures Appl. **25**, 239–241 (1980)

[HuMo75] R.E. Huff, P.D. Morris, Dual spaces with the Krein–Milman property have the Radon–Nikodým property. Proc. Am. Math. Soc. **49**, 104–108 (1975)

[Ja88] K. Jarosz, Any Banach space has an equivalent norm with trivial isometries. Isr. J. Math. **64**, 49–56 (1988)

[JM97] M. Jiménez-Sevilla, J.P. Moreno, Renorming Banach paces with the Mazur intersection property. J. Funct. Anal. **144**, 486–504 (1997)

[JoWo79] J. Johnson, J. Wolfe, Norm attaining operators. Stud. Math. **65**, 7–19 (1979)

[Jo76] W.B. Johnson, A reflexive Banach space which is not sufficiently Euclidean. Stud. Math. **55**(2), 201–205 (1976)

[JoLin74] W.B. Johnson, J. Lindenstrauss, Some remarks on weakly compactly generated Banach spaces. Isr. J. Math. **17**, 219–230 (1974)

[JoLin01] W.B. Johnson, J. Lindenstrauss (eds.), *Handbook of the Geometry of Banach Spaces I* (Elsevier, Amsterdam, 2001)

[JoLin01b] W.B. Johnson, J. Lindenstrauss, *Basic Concepts in the Geometry of Banach Spaces*, ed. by W.B. Johnson, J. Lindenstrauss. Handbook of the Geometry of Banach Spaces I (Elsevier, Amsterdam, 2001)

[JoLin03] W.B. Johnson, J. Lindenstrauss (eds.), *Handbook of the Geometry of Banach Spaces II* (Elsevier, Amsterdam, 2003)

[JLPS02] W.B. Johnson, J. Lindenstrauss, D. Preiss, G. Schechtman, Lipschitz quotients from metric trees and from Banach spaces containing ℓ_1. J. Funct. Anal. **194**, 332–346 (2002)

[JLS96] W.B. Johnson, J. Lindenstrauss, G. Schechtman, Banach spaces determined by their uniform structures. Geom. Funct. Anal. (GAFA) **6**, 430–470 (1996)

[JoOd05] W.B. Johnson, E. Odell, The diameter of the isomophism class of a Banach space. Ann. Math. **162**(1), 423–437 (2005)

[JoSch14] W.B. Johnson, G. Schechtman, Subspaces of L_p that embed into $L_p(\mu)$ with μ finite. Isr. J. Math. **203**, 211–222 (2014)

[JoSz12] W.B. Johnson, A. Szankowski, Hereditary approximation property. Ann. Math. **176**, 1987–2001 (2012)

[JoSz14] W.B. Johnson, A. Szankowski, The trace formula in Banach spaces. Isr. J. Math. **203**, 389–404 (2014)

[JoZi89] W.B. Johnson, M. Zippin, Extension of operators from subspaces of $c_0(\Gamma)$ into $C(K)$ spaces. Proc. Am. Math. Soc. **107**(3), 751–754 (1989)

[Ka67] M.I. Kadets, A proof of topological equivalence of all separable infinite-dimensional Banach spaces. (Russian), Funk. Anal. i Prilozen **1**, 61–70 (1967)

[KaMar12] V. Kadets, M. Martín, Extension of isometries between unit spheres of finite-dimensional polyhedral Banach spaces. J. Math. Anal. Appl. **396**, 441–447 (2012)

[KMP00] V. Kadets, M. Martín, R. Payá, Recent progress and open questions on the numerical index of Banach spaces. Rev. R. Acad. Cienc. Exactas Fís. Natl. Ser. A Math. RACSAM **100**(1–2), 155–182 (2006)

[Kal65] N.J. Kalton, The basic sequence problem. Stud. Math. **116**, 167–187 (1965)

[Kal74] N.J. Kalton, Spaces of compact operators. Math. Ann. **208**, 267–278 (1974)

[Kal03] N.J. Kalton, *Quasi-Banach Spaces*, ed. by W.B. Johnson, J. Lindenstrauss. Handbook of the Geometry of Banach Spaces II (Elsevier, Amsterdam, 2003), pp. 1099–1130

[Kal04] N.J. Kalton, Spaces of Lipschitz and Hölder functions and their applications. Collect. Math. **55**(2), 171–217 (2004)

[Kal06] N.J. Kalton, Extension problems for $C(K)$ spaces and twisted sums, in *Methods in Banach Space Theory*, ed. by J.M.F. Castillo, W.B. Johnson. London Mathematical Society Lecture Notes Series, vol. 337 (Cambridge University Press, Cambridge, 2006)

[Kal07] N.J. Kalton, Coarse and uniform embeddings into reflexive spaces. Q. J. Math. **58**, 393–414 (2007)

[Kal08] N.J. Kalton, The nonlinear geometry of Banach spaces. Rev. Mat. Complut. **21**, 7–60 (2008)

[Kal11] N.J. Kalton, Lipschitz and uniform embeddings into ℓ_∞. Fund. Math. **212**, 53–69 (2011)

[Kal12] N.J. Kalton, The uniform structure of Banach spaces. Math. Ann. **354**, 1247–2888 (2012)

[Kau91] R. Kaufman, Topics on analytic sets. Fundam. Math. **139**, 215–229 (1991)

[Kauf15] P.L. Kaufmann, Products of Lipschitz-free spaces and applications. Stud. Math. **226**(3), 213–227 (2015)

[Kec94] A.S. Kechris, *Classical Descriptive Set Theory*. Graduate Texts in Mathematics, vol. 156 (Springer, Berlin, 1994)

[KMS01] P.S. Kenderov, W.B. Moors, S. Sciffer, A weak Asplund space whose dual is not weak* fragmentable. Proc. Am. Math. Soc. **192**, 3741–3757 (2001)

[Ki65] W.A. Kirk, A fixed point theorem for mappings which do not increase distances. Am. Math. Mon. **72**, 1004–1006 (1965)

[Ki95] W.A. Kirk, Some questions in metriuc fixed point theory. *Recent Advances on Metric Fixed Point Theory (Sevilla, 1995)*. Ciencias, vol. 48 (University of Sevilla, Sevilla, 1996), pp. 73–97

[Klee59] V. Klee, Some new results on smoothness and rotundity in normed linear spaces. Math. Ann. **139**, 51–63 (1959)

[Kom94] R. Komorowski, On constructing Banach spaces with no unconditional bases. Proc. Am. Math. Soc. **120**(1), 101–107 (1994)

[KoTo95,98] R. Komorowski, N. Tomczak-Jaegermann, Banach spaces without local unconditional structure. Isr. J. Math. **89**, 205–226 (1995); ibidem **105**, 85–92 (1998)

[Kosz04] P. Koszmider, Banach paces of continuous functions with few operators. Math. Ann. **330**(1), 151–183 (2004)

[Kosz05] P. Koszmider, A space $C(K)$ where all nontrivial complemented subspaces have big densities. Stud. Math. **186**(2), 109–127 (2005)

[Kosz06] P. Koszmider, On decompositions of Banach spaces of continuous functions on Mrowka's spaces. Proc. Am. Math. Soc. **133**(7), 2137–2146 (2006)

[Kosz10] P. Koszmider, A survey on Banach spaces $C(K)$ with few operators. Rev. R. Acad. Cienc. Exactas Fís. Natl. Ser. A Math. RACSAM **104**(2), 309–326 (2010)

[Kosz13] P. Koszmider, On large indecomposable Banach spaces. J. Funct. Anal. **246**, 1779–1805 (2013)

[Kosz] P. Koszmider, Uncountable equilateral sets in Banach spaces of the form $C(K)$, to appear

[KMM09] P. Koszmider, M. Martín, J. Merí. Extremely non-complex $C(K)$ spaces. J. Math. Anal. Appl. **350**, 601–615 (2009)

[Ko75] C. Kottmann, Subsets of the unit ball that are separated by more than one. Stud. Math. **53**, 15–27 (1975)

[Kun81] K. Kunen, A compact L-space under CH. Topol. Appl. **12**, 283–287 (1981)

[Kur11] O. Kurka, Structure of the set of norm attaining functionals on strictly convex spaces. Can. Math Bull. **54**(2), 302–310 (2011)

[La04] S. Lajara, ALUR dual renormings of Banach spaces. J. Math. Anal. Appl. **299**, 221–226 (2004)

[La11] S. Lajara, Average locally uniform rotundity and a class of nonlinear maps. Nonlinear Anal. **74**, 1937–1944 (2011)

[Li08] P.-K. Lin, There is an equivalent nom on ℓ_1 that has the fixed point property. Nonlinear Anal. **68**, 2303–2308 (2008)

[Lin63] J. Lindenstrauss, On operators which attain their norm. Isr. J. Math. **3**, 139–148 (1963)

[Lin70] J. Lindenstrauss, Some aspects of the theory of Banach spaces. Adv. Math. **5** (1970), 159–180.

[Lin72] J. Lindenstrauss, Weakly compact sets, their topological properties and spaces they generate. Ann. Math. Stud. **69** (1972)

[LPT10] J. Lindenstrauss, D. Preiss, J. Tišer, Fréchet differentiability of Lipschitz functions via variational principle. J. Eur. Math. Soc. **12**, 385–412 (2010)

[LPT12] J. Lindenstrauss, D. Preiss, J. Tišer, *Fréchet Differentiability of Lipschitz Functions and Porous Sets in Banach Spaces*. Annals of Mathematics Studies, vol. 179 (Princeton University Press, Princeton 2012)

[LinTza71] J. Lindenstrauss, L. Tzafriri, On the complemented subspace problem. Isr. J. Math. **9**, 263–269 (1971)

[LinTza73] J. Lindenstrauss, L. Tzafriri, *Classical Banach Spaces*. Lecture Notes in Mathematics, vol. 338 (Springer, Berlin, 1973)

[LinTza77] J. Lindenstrauss, L. Tzafriri, *Classical Banach Spaces I. Sequence Spaces* (Springer, Berlin, 1977)

[Lo00] V. Lomonosov, A counterexample to the Bishop–Phelps theorem in complex spaces. Isr. J. Math. **115**, 25–28 (2000)

[LAТo09] J. López-Abad, S. Todorčević, A c_0-saturated Banach space with no long unconditional basic sequence. Trans. Am. Math. Soc. **361**, 4541–4560 (2009)

[Lov55] A. Lovaglia, Locally uniformly convex spaces. Trans. Am. Math. Soc. **78**, 225–238 (1955)

[Lus85] W. Lusky, A note on Banach spaces containing c_0 or C_∞. J. Funct. Anal. **62**, 1–7 (1985)

[Lus98] W. Lusky, Three space properties and basis extensions. Isr. J. Math. **107**, 17–27 (1998)

[Man72] P. Mankiewicz, On extension of isometries in normed linear spaces. Bull. Acad. Polon. Sci. Sér. Sci. Math. Astronom. Phys. **20**, 367–371 (1972)

[Man73] P. Mankiewicz, On the differentiability of Lipschitz mappings in Fréchet spaces. Stud. Math. **45**, 15–29 (1973)

[MaPl10] W. Marciszewski, G. Plebanek, On Corson compacta and embeddings of $C(K)$ spaces. Proc. Am. Math. Soc. **138**(12), 4281–4289 (2010)

[Mar14] M. Martín, Norm-attaining compact operators. J. Funct. Anal. **267**, 1585–1592 (2014)

[MMOT10] J.F. Martínez, A. Moltó, J. Orihuela, S.L. Troyanski, On locally uniformly rotund renormings in C(K) spaces. Can. J. Math. **62**(3), 595–613 (2010)

[MaPl05] V.K. Maslyuchenko, A.M. Plichko, Some open problems in functional analysis and function theory. Extracta Math. **20**(1), 51–70 (2005)

[Mat97] E. Matoušková, Convexity of Haar null sets. Proc. Am. Math. Soc. **125**, 1793–1799 (1997)

[Mat99] E. Matoušková, An almost nowhere Fréchet smooth norm on superreflexive spaces. Stud. Math. **133**(1), 93–99 (1999)

[MatSt96] E. Matoušková, C. Stegall, A characterization of reflexive Banach spaces. Proc. Am. Math. Soc. **124**, 1083–1090 (1996)

[Mau81] B. Maurey, *Points Fixes des Contractions de Certains Faiblement Compacts de L^1*. Semin. Anal. Fonct., 1980–1981 (École Polytechnique, Palaiseau, 1981)

[Mau03] B. Maurey, *Banach Spaces with Few Operators*, ed. by W.B. Johnson, J. Lindenstrauss. Handbook of the Geometry of Banach Spaces II (Elsevier, Amsterdam, 2003), pp. 1247–1332

[Maz33] S. Mazur, Über schwache Konvergentz in den Raumen L^p. Stud. Math. **4**, 128–133 (1933)

[Megg98] R.E. Megginson, *An Introduction to Banach Space Theory*. Graduate Texts in Mathematics, vol. 183 (Springer, Berlin, 1998)

[MerSta06] S.K. Mercourakis, E. Stamati, A new class of weakly K-analytic Banach spaces. Comment. Math. Univ. Carol. **47**(2), 291–312 (2006)

[MerVa14] S.K. Mercourakis, G. Vassiliadis, Equilateral sets in infinite-dimensional Banach spaces. Proc. Am. Math. Soc. **142**, 205–212 (2014)

[Mil72] V.D. Milman, Geometric theory of Banach spaces. II: Geometry of the unit sphere. Russ. Math. Surv. **26**, 79–163 (1972)

[Min70] G.J. Minty, On the extension of Lipschitz, Lipschitz–Hölder continuous and monotone functions. Bull. Am. Math. Soc. **76**, 334–339 (1970)

[MOTV09] A. Moltó, J. Orihuela, S.L. Troyanski, M. Valdivia, *A Non-Linear Transfer Technique for Renorming*. Lecture Notes in Mathematics, vol. 1951 (Springer, Berlin, 2009)

[MZZ15] V. Montesinos, P. Zizler, V. Zizler, *An Introduction to Modern Analysis* (Springer, Berlin, 2015)

[MoSu06] W.B. Moors, S. Somasundaram, A Gâteaux differentiability space that is not weak Asplund. Proc. Am. Math. Soc. **134**, 2745–2754 (2006)

[Mo] P. Morris, Disapperance of extreme points. Proc. Am. Math. Soc. **88**(2), 244–246 (1983)

[Muj77] J. Mujica, Separable quotients of Banach spaces. Rev. Math. **10**, 299–330 (1977)

[NaRa] A. Naor, Y. Rabani, On Lipschitz extension from finite subsets, to appear

[NaSch07] A. Naor, G. Schechtman, Planar earthmover is not in L_1. SIAM J. Comput. **37**(3), 804–826 (2007)

[OdSc94] E. Odell, Th. Schlumprecht, The distortion problem. Acta Math. **173**, 259–281 (1994)

[OdSc01] E. Odell, Th. Schlumprecht, *Distortion and Asymptotic Structure*, ed. by W.B. Johnson, J. Lindenstrauss. Handbook of the Geometry of Banach Spaces II (Elsevier, Amsterdam, 2003), pp. 1333–1360

[Oja08] E. Oja, The strong approximation property. J. Math. Anal. Appl. **338**, 407–415 (2008)

[OST] J. Orihuela, R.J. Smith, S.L. Troyanski, Strictly convex norms and topology. Proc. Lond. Math. Soc. **104**(3), 197–222 (2012)

[Par81] J.R. Partington, Subspaces of certain Banach sequence spaces. Bull. Lond. Math. Soc. **13**(2), 163–166 (1981)

[Pe94] J. Pelant, Embedding into c_0. Topol. Appl. **57**, 259–269 (1994)

[PHK06] J. Pelant, P. Holický, O.F.K. Kalenda, $C(K)$ spaces which cannot be uniformly embedded into $c_0(\Gamma)$. Fund. Math. **192**, 245–254 (2006)

[Pe57] A. Pełczyński, A property of multilinear operations. Stud. Math. **16**(2), 173–182 (1957)

[Pe06] A. Pełczyński, Selected problems on the structure of complemented subspaces of Banach spaces, in *Methods in Banach Space Theory*. London Mathematical Society Lecture Note Series, vol. 337 (Cambridge University Press, Cambridge, 2006), pp. 341–354

[PeBe79] A. Pełczyński, C. Bessaga, *Some Aspect of the Present Theory of Banach Spaces* (S. Banach: Travaux sur l'Analyse Fonctionnelle, Warszaw, 1979)

[PeSi64] A. Pełczyński, I. Singer, On non-equivalent bases and conditional bases in Banach spaces. Stud. Math. **25**, 5–25 (1964)

[Ph57] R.R. Phelps, Subreflexive normed linear spaces. Arch. Math. **8**, 444–450 (1957)

[Ph60] R.R. Phelps, A representation theorem for bounded convex sets. Proc. Am. Math. Soc. **11**, 976–983 (1960)

[Piet09] A. Pietsch, Long-standing open problems of Banach space theory. Quaest. Math. **32**, 321–327 (2009)

[Pis85] G. Pisier, *Factorization of Linear Operators and Geometry of Banach Spaces*. CBMS Regional Conference Series in Mathematics, vol. 60 (American Mathematical Society, Providence, 1985)

[Pis88] G. Pisier, Weak Hilbert spaces. Proc. Lond. Math. Soc. **56**, 547–579 (1988)

[Ple04] G. Plebanek, A construction of a Banach space $C(K)$ with few operators. Topol. Appl. **143**, 217–239 (2004)

[PleSo15] G. Plebanek, D. Sobota, Countable tightness in spaces of regular probability measures. Fundam. Math. **229**(2), 159–170 (2015)

[PliYo00] A. Plichko, D. Yost, Complemented and uncomplemented subspaces of Banach spaces. Extracta Math. **15**, 335–371 (2000)

[PliYo01] A. Plichko, D. Yost, The Radon–Nikodým property does not imply the separable complementation property. J. Funct. Anal. **180**, 481–487 (2001)

[Pr90] D. Preiss, Differentiability of Lipschitz functions on Banach spaces. J. Funct. Anal. **91**, 312–345 (1990)

[Pr10] D. Preiss, Tilings of Hilbert spaces. Mathematika **56**, 217–230 (2010)

[PrTi95] D. Preiss, J. Tišer, Two unexpected examples concerning differentiability of Lipschitz functions in Banach spaces, in *Geometric Aspects of Functional Analysis (Israel, 1992–1994)*, ed. by J. Lindenstrauss, V. Milman. Operator Theory: Advances and Applications, vol. 77 (Birkhäuser, Boston, 1995), pp. 219–235

[PrZa84] D. Preiss, L. Zajíček, Fréchet differentiation of convex functions in a Banach space with a separable dual. Proc. Am. Math. Soc. **91**, 202–204 (1984)

[Re88] C.J. Read, The invariant subspace problem for a class of Banach spaces. II. Hypercyclic operators. Isr. J. Math. **63**(1), 1–40 (1988)

[Ri84] M. Ribe, Existence of separable uniformly homeomorphic nonisomorphic Banach spaces. Isr. J. Math. **48**, 139–147 (1984)

[Rm] M. Rmoutil, Norm-attaining functionals and proximinal subspaces, to appear

[Rol69] S. Rolewicz, On orbit elements. Stud. Math. **32**, 17–22 (1969)

[Ro74] H.P. Rosenthal, The heredity problem for weakly compactly generated Banach spaces. Compos. Math. **28**(1), 83–111 (1974)

[Ro03] H.P. Rosenthal, *The Banach Space $C(K)$*, ed. by W.B. Johnson, J. Lindenstrauss. Handbook of the Geometry of Banach Spaces II (Elsevier, Amsterdam, 2003), pp. 1549–1602

[Sch83] W. Schachermayer, Norm Attaining Operators on some Classical Banach spaces. Pac. J. Math. **105**, 427–438 (1983)

[Sch87] W. Schachermayer, The Radon–Nikodým property and the Krein-Milman property are equivalent for strongly regular sets. Trans. Am. Math. Soc. **303**(2), 673–687 (1987)

[Sch87b] W. Schachermayer, Some more remarkable properties of the James-tree space. Contemp. Math. **85**, 465–496 (1987)

[SSW89] W. Schachermayer, A. Sersouri, E. Werner, Moduli of non-dentability and the Radon–Nikodým property in Banach spaces. Isr. J. Math. **65**, 225–257 (1989)

[Schl91] Th. Schlumprecht, An arbitrary distortable Banach space. Isr. J. Math. **76**, 81–95 (1991)

[Sin78] I. Singer, On the distance of nonreflexive spaces to the collection of all conjugate spaces. Bull. Aust. Math. Soc. **18**, 461–464 (1978)

[SimYo89] B. Sims, D. Yost, Linear Hahn-Banach extension properties. Proc. Edinb. Math. Soc. **32**, 53–57 (1989)

[Sm81] M.A. Smith, A Banach space that is MLUR but not HR. Math. Ann. **256**, 277–279 (1981)

[Smi07] R.J. Smith, Trees, Gâteaux norms and a problem of Haydon. J. Lond. Math. Soc. **76**(2), 633–646 (2007)

[SmiTr10] R.J. Smith, S.L. Troyanski, Renormings of $C(K)$ spaces. Rev. R. Acad. Cienc. Exactas Fís. Natl. Ser. A Math. RACSAM **104**(2), 375–412 (2010)

[Ta85] M. Talagrand, Espaces de Baire et espaces de Namioka. Math. Ann. **270**(2), 159–164 (1985)

[Ta86] M. Talagrand, Renormages de quelques $C(K)$. Isr. J. Math. **54**, 327–334 (1986)

[Tana14] R. Tanaka, A further property of spherical isometries. Bull. Aust. Math. Soc. **90**, 304–310 (2014)

[Tan96] W.-K. Tang, A note on preserved smoothness. Serdica Math. J. **22**, 29–32 (1996)

[Te87] P. Terenzi, Successioni regolari negli spazi di Banach. Milan J. Math. **57**(1), 275–285 (1987)

[Tin87] D. Tingley, Isometries of the unit sphere. Geom. Dedicata **22**, 371–378 (1987)

[To97] S. Todorčević, *Topics in Topology*. Lecture Notes in Mathematics, vol. 1652 (Springer, Berlin, 1997)

[To06] S. Todorčević, Biorthogonal systems and quotient spaces via Baire category theory. Math. Ann. **335**, 687–715 (2006)

[Tor81] H. Torunczyk, Characterizing Hilbert space topology. Fund. Math. **111**, 247–262 (1981)

[Tro67] S.L. Troyanski, On topological equivalence of spaces $c_0(\aleph)$ and $\ell_1(\aleph)$. Bull. Acad. Polon. Sci. Ser. Sci. Math. Astr. Phys. **15**, 389–396 (1967)

[Tro70] S.L. Troyanski, An example of a smooth space, the dual of which is not strictly convex. Stud. Math. **35**, 305–309 (1970) (in Russian)

[Tro90] S.L. Troyanski, Gâteaux differentiable norms in L_p. Math. Ann. **287**, 221–227 (1990)

[Tsi74] B.S. Tsirelson, Not every Banach pace contains ℓ_p or c_0. Funct. Anal. Appl. **8**, 138–141 (1974)

[Van92] J. Vanderwerff, Fréchet differentiable norms on spaces of countable dimension. Arch. Math. **58**, 471–476 (1992)

[Van93] J. Vanderwerff, Second order Gâteaux differentiability and an isomorphic characterization of Hilbert spaces. Q. J. Math. **44**(2), 249–255 (1993)

[Vas81] L. Vašák, On a generalization of weakly compactly generated Banach spaces. Stud. Math. **70**, 11–19 (1981)

[Vla69] L.P. Vlasov, Approximative properties of sets in normed linear spaces. Russ. Math. Surv. **28**, 1–66 (1973)

[Vla70] L.P. Vlasov, Almost convexity and Chebyshev sets. Math. Notes Acad. Sci. URSS **8**, 776–779 (1970)

[Wa01] H.M. Wark, A nonseparable reflexive Banach space on which there are few operators. J. Lond. Math. Soc. **64**(2), 675–689 (2001)

[Wea99] N. Weaver, *Lipschitz Algebras* (World Scientific, Singapore, 1999)

[Wer01] D. Werner, Recent progress on the Daudavet property. Ir. Math. Soc. Bull. **46**, 77–97 (2001)

[Woj91] P. Wojtaszczyk, *Banach Spaces for Analysts*. Cambridge Studies in Advanced Mathematics, vol. 25 (Cambridge University Press, Cambridge, 1991)

[Zip03] M. Zippin, *Extension of Bounded Linear Operators*, ed. by W.B. Johnson, J. Lindenstrauss. Handbook of the Geometry of Banach Spaces II (Elsevier, Amsterdam, 2003), pp. 1703–1742

[Ziz71] V. Zizler, On some rotundity and smoothness properties of Banach spaces. Diss. Math. **87**, 1–33 (1971)

[Ziz73] V. Zizler, On some extremal problems in Banach spaces. Math. Scand. **32**, 214–224 (1973)

[Ziz86] V. Zizler, Renormings concerning Mazur's intersection property of balls for weakly compact convex sets. Math. Ann. **276**(1), 61–66 (1986)

[Ziz03] V. Zizler, *Nonseparable Banach Spaces*, ed. by W.B. Johnson, J. Lindenstrauss. Handbook of the Geometry of Banach Spaces II (Elsevier, Amsterdam, 2003), pp. 1743–1816

List of Concepts and Problems

Concept	Problem number
Baire space	285
Basis, Auerbach	113, 294
Basis constant	3
Basis, Markushevich	237
Basis, Markushevich, norming	111, 112
Basis, of a topology	286
Basis, Schauder	1, 6, 12, 18, 19, 108, 109, 193, 230, 231
Basis, Schauder, equivalent	1, 108
Basis, Schauder, long	9
Basis, Schauder, monotone	109
Basis, Schauder, unconditional	1, 5, 6, 7, 8, 12, 20, 108, 137, 252, 294
Basis set	288, 289
Bilinear form	122
Bump, Fréchet	146, 175, 187, 264
Bump, Gâteaux	130, 134, 179, 221
Bump, Lipschitz	146, 182, 183, 221
Bump, C^1	144, 145, 188
Bump, C^2	10, 22
Bump, C^k	174, 180, 182, 183
Bump, C^∞	2, 136, 180
Bump, loc. depend. finitely many coordinates	222, 225
Bump, twice differentiable	166

© Springer International Publishing Switzerland 2016
A.J. Guirao et al., *Open Problems in the Geometry and Analysis of Banach Spaces*,
DOI 10.1007/978-3-319-33572-8

Concept	Problem number
Function, nonexpansive	297
Function, real analytic	171, 178
Function, Riemann integrable	302
Function, separately continuous	285
Function, uniformly continuous	199
Functional, norm-attaining	204, 206, 208, 217
Functional, support	202, 203, 207
H^∞	53
Hilbert cube	99, 269
Homeomorphism, Lipschitz	239, 240, 241, 242, 243, 244, 245, 246, 247, 251, 252, 253, 254, 255, 256, 257, 258, 259, 260
Homeomorphism, uniform	250, 272, 273, 274, 275, 276, 278, 279, 280, 281, 282
Homeomorphic to Hilbert	268, 270
Homeomorphism, weak	271
Isomorphism	239, 240, 243, 244, 245, 249
Isometry	83, 84, 85, 86, 87, 96, 99
James tree space	265
ℓ_1	30, 58, 100, 129, 148, 149, 150, 152, 223, 233, 235, 243, 246, 265, 287
ℓ_2	1, 42, 77, 79, 124, 129, 172, 178, 233, 251, 275, 277
ℓ_3	179
ℓ_p	4, 78, 176, 186, 262, 291
ℓ_∞	58, 92, 107, 237, 238, 245, 248, 252, 253, 261, 287
L_1	30, 85, 90, 234, 275, 288
L_p	8, 9, 90, 280
Linearly isometric to Hilbert	84, 85
Linearly isomorphic to Hilbert	22, 254
Linearly isomorphic to hyperplanes	2
Lipschitz embedding	248, 249, 261

Concept	Problem number
Space, superreflexive	66, 127, 163, 164, 167, 239, 256, 257, 273, 295, 297
Space, weak Asplund	218, 219, 220, 266
Space, weak Hilbert	18, 19, 20, 21, 22
Space, WCG	69, 93, 111, 112, 114, 116, 117, 141, 169, 244, 270
Space, WLD	134, 135, 138
Space, weakly sequentially complete	282
Subspace, complemented	1, 6, 23, 24, 25, 30, 31, 33, 34, 35, 36, 37, 38, 39, 83
Subspace, infinite increasing sequence	32
Subspace, invariant	42, 43
Subspace, norming	117, 121
subspace, proper	89
Subspace, quasicomplemented	32
Tiling	81, 82
Tsirelson space	292
Uniform normal structure	295
Weak compactness	287

Symbol Index

A

$\aleph$, a cardinal number, viii

$\aleph_0$, the cardinal number of $\mathbb{N}$, viii

$\aleph_1$, the first uncountable cardinal, viii

α, an ordinal number, viii

ALUR, average locally uniformly rotund, 49

AP, the approximation property, 10

B

BAP, the bounded approximation property, 26

bc(X), the basis constant of X, 5

$B_\varepsilon(x)$, the closed ball of radius ε centered at x, 115

$BV(\mathbb{R}^n)$, real-valued functions with bounded variation on $\mathbb{R}^n$, 27

B_X, the closed unit ball of X, viii

$B(X, Y)$, the space of all bounded operators from X into Y, 95

C

$\mathbb{C}$, the field of complex numbers, 12

c, the cardinality of $\mathbb{R}$, viii

c_0, the space of all null sequences, 3

$c_0(\Gamma)$, the space of all null Γ-sequences, 55

CCC, *see* compact, CCC

CH, the continuum hypothesis, 86

$C(K)$, the space of continuous functions on K, 3

CPCP, the convex point of continuity property, 62

C^*PCP, the convex* point of continuity property, 62

D

D, a diagonal operator, 11

$\delta(\varepsilon)$, the modulus of convexity, 3, 89

δ_t, the Dirac delta at t, 101

diam(M), the diameter of a set M, 87

dim (X), the dimension of a space X, 45

D_φ, the Gâteaux differential operator at φ, 113

$d(X, Y)$, the Banach–Mazur distance, 5

E

$e_n(T)$, the n-th entropy number of T, 34

F

FDD, a finite-dimensional decomposition, 29

$\mathcal{F}(M)$, the Lipschitz-free space over M, 104

FPP, Fixed Point Property, 130

H

HI, hereditarily indecomposable, 3

I

Id_X, the identity operator on X, 9

J

JL_0, the space of Johnson and Lindenstrauss, 114

$J(\omega_1)$, the long James space, 64

JT, the James tree space, 71

© Springer International Publishing Switzerland 2016

A.J. Guirao et al., *Open Problems in the Geometry and Analysis of Banach Spaces*, DOI 10.1007/978-3-319-33572-8

Subject Index

Printed in the United States
By Bookmasters